CW00473353

Understanding Forces

How forces work

Studymates

Many other titles in preparation

Understanding Forces
How Forces Work

Barry Dawson

www.**studymates**.co.uk

© 2003 by Barry Dawson

ISBN 1 84285 011 3

First published in 2003 by Studymates Limited, PO Box 2, Taunton, Somerset TA3 3YE, United Kingdom

Telephone: (01823) 432002
Fax: (01823) 430097
Website: http://www.studymates.co.uk

All rights reserved. No part of this work may be reproduced or stored in an information retrieval system without the express permission of the Publishers given in writing.

Note: The contents of this book are offered for the purposes of general guidance only and no liability can be accepted for any loss or expense incurred as a result of relying in particular circumstances on statements made in this book. Readers are advised to check the current position with the appropriate authorities before entering into personal arrangements.

Typeset by PDQ Typesetting, Newcastles-under-Lyme.
Printed in the UK by The Baskerville Press Ltd, Salisbury - www.baskervillepress.com

Contents

Preface

Physics is a subject which has a few recurring themes throughout all the different branches. Forces is one and energy is another. This book therefore looks at each of the branches of physics where forces are important. These forces can be very large, such as the force which holds the Earth in orbit round the sun, or much smaller such as the force exerted on a fly's leg as it crawls over a surface. There are many physics textbooks in print which deal with the whole course, but I know of no other book which just considers this aspect. Many of the existing books are quite formal in their approach and this can be quite daunting to students who perhaps lack confidence. This book has a much more informal style. I want you to think that it is your book which was written especially for you. Physics should be something that you feel part of, not something which you keep at arm's length.

This book is aimed at two different groups:

▶ students who are preparing for A level physics and find that they need a less formal or more personal explanation

▶ students who are at university studying a science-related topic and need to brush up A level work on forces, or who perhaps did not study physics.

A useful feature is the collection of web sites at the end of the book. Some of these have been chosen because they give a full account of important topics and will therefore provide yet another source to help your understanding. Others have been chosen just because they are fun. One worth particular mention is the site which allows you to design your own roller coaster and to see whether it fails to get up the second hill, flies off the second hill or fails to get round the loop.

This book originated as the result of a chance encounter. My second interest after physics is information technology, indeed I spend as much time teaching that subject as I do teaching physics. It was at an A level standardisation meeting for information technology in November 1999 that I started speaking to a colleague named Graham Lawlor whom I had not met before. From our conversation he discovered that I had been preparing students for A level physics for many years and he asked if I would write a book about forces. He is therefore the person who I must thank for getting this project started.

My artistic skills are poor and I must thank Helen J Dawson for using her superior talents to create the better drawings such as Figures 9, 11 to 17, 19, 21, 23 and 24. Her small people can brighten even the most difficult of concepts.

Andrew J Dawson, who has recently finished an A level course himself, was kind enough to proof read the final manuscript and to suggest ways of making the text easier to understand.

Finally I must thank my wife Jan Dawson who has been left alone with our dog Susie while I have been working at the computer. Jan always comes to my rescue when I am in trouble over spelling or grammar.

1

An Introduction to Forces

One-minute summary – The study of forces and their effects appears in many different branches of physics. Here we will find out the different effects which forces can have, and we will see how the same basic rules can be applied. For example a force keeps the earth in place as it goes round the sun, a much smaller force is used when you close the car door, and an even smaller force is used to press your pencil onto the paper so that it leaves a graphite mark behind. In this chapter we will:

▶ learn what is meant by a force
▶ understand the terms 'mass' and 'weight' as used in physics
▶ find out about the stretching of springs
▶ learn how a forcemeter works.

What is a force?

Put as simply as possible, a force is a push or a pull. When the muscles in our body contract, it is because they are exerting a force. Muscles can only exert a force when they contract so they can only pull. The bone to which they are attached can push or pull, according to how the muscle is attached. Forces can have different effects – they can stretch things, compress things, change speed, change direction, twist or just balance out other forces so that nothing happens. Make yourself a chart giving a few everyday examples for each of these.

Force and weight

Let us think for a moment about a burger. It could be a beef burger or a veggie burger, depending on your taste, and has a mass of about 100g. Its **mass** is a measure of the amount of material that it contains. It can also be thought of as a measure of the number of particles in it, although being organic its particles have a rather complicated structure. The burger will provide us with a certain amount of nourishment. A burger with twice the mass would provide us with twice the nourishment. On earth, gravity exerts a force of about 10N (10 newtons or more exactly 9.81N) on an object whose mass equals one kilogram, so the force on our burger should be about one newton. This force of gravity on our burger is called its **weight**.

If we now take our burger to the moon where **gravity** is much less than that on Earth it will contain just as many particles and will give us just as much nourishment as when we eat it on Earth. Its mass is still 100g. Its weight however is less than 0.2N because gravity is less. This means that on the moon the mass of the burger is the same as on Earth but its weight is less, due to gravity being lower. In deep space our burger still has a mass of 100g but its weight will be zero. (There is more information about gravity in Chapter 7.)

Students learn better if they stop and perform some related task after each small chunk of work. Perhaps this would be a good time to cook a burger and find out how much nourishment you get from 100g.

Hooke's law

Robert Hooke was born in 1635. He was curator of experiments at the Royal Society. He was doing his work at the same time as Robert Boyle and Isaac Newton. One of Hooke's investigations resulted in **Hooke's law** which states:

**The extension of a spiral spring is proportional
to the force applied.**

This law applies for the first part of the extension but eventually the **elastic limit** is reached. The law also applies to springs which can be compressed (such as a bed spring). This law can be stated as $F = kx$

where F is the applied force in newtons, x is the extension in metres and k is known as the spring constant. The units for spring constant are Nm^{-1} (which is the same as newtons per metre or N/m).

The spring constant is the force needed to stretch the spring by unit length.

Unit length in this case is one metre but in practice we are unlikely to cause such a large extension.

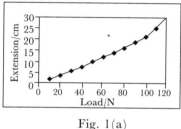

Fig. 1(a)

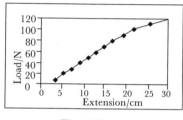

Fig 1(b)

Two graphs are shown above. Figure 1(a) shows extension against load and Figure 1(b) shows load against extension. On both graphs the straight-line part shows the region over which Hooke's law is obeyed. In other words, the region for which the extension is proportional to the force applied. If you were to carry out this experiment you would apply different forces by hanging masses on the spring and measuring the resulting extension. By convention you would plot the quantity which you change (the force) on the horizontal axis and the quantity which you measure (the extension) on the vertical axis. In this case the line curves upwards after the elastic limit has been reached because the increases in length are produced by relatively smaller increases in the force. This is shown in Figure 1(a). The second graph is sometimes used because the gradient of the straight-line section is equal to the spring constant k. The **elastic limit** is so called because after this the spring no longer returns to its original length when the force is removed.

Exercise 1.1

Force N	0	1.00	2.00	3.00	4.00	5.00	6.00
Extension cm	0	0.45	0.85	1.27	1.69	2.10	2.55

Force N	7.00	8.00	9.00	10.00	11.00
Extension cm	2.95	3.55	4.25	5.10	6.15

These results come from an experiment where a number of different forces have been applied to a spiral spring and the corresponding extensions have been measured. Plot a graph with the axes as in Figure 1(b) and from it find the spring constant and the elastic limit of the spring. Remember that you should draw the best-fit line or curve. You must not join the points with a series of short straight lines.

Example
During an experiment when a spring was being stretched always within its elastic limit, the following results were recorded. When a force of 12N was applied the total length was 0.47m and when the force was increased to 22N the total length was 0.72m. Find:

(a)   the spring constant k

(b)   the original length of the spring before any force was applied and

(c)   the length of the spring when the force applied is 25N.

Give the assumption which you have made in part (c).

We do not know the original length of the spring so we cannot work out the extension but we do know that an increase of force equal to $22N - 12N$ caused an increase in extension of $0.72m - 0.47m$.

(a)
$$F = kx$$
$$k = \frac{F}{x}$$

$$k = \frac{\text{increase in force}}{\text{increase in extension}}$$

$$= \frac{22 - 12}{0.72 - 0.47}$$

$$= \frac{10}{0.25}$$

$$= 40\text{Nm}^{-1}$$

(b) A force of 12N made the total length of the spring 0.47m. We can use the spring constant to find out how much of this was extension.

$$F = kx$$

$$x = \frac{F}{k}$$

$$= \frac{12}{40}$$

$$= 0.30m$$

The original length of the spring was therefore 0.47m − 0.30m = 0.17m.

(c) We need to assume that Hooke's law still applies at a load of 25N. In other words the elastic limit has not been reached. We first need to find the extension at 25N and then we must add the original length.

$$F = kx$$

$$x = \frac{F}{k}$$

$$= \frac{25}{40}$$

$$= 0.625m$$

The total length would therefore be 0.625 + 0.17 = 0.795 = 0.80m.

We correct the final answer to two significant figures since it would not be sensible to suggest that the answer is correct to three significant figures when we are only given figures which are correct to two significant figures in the question.

Note: If the figures in the question had been 12.0, 22.0, 0.470 and 0.720 instead of 12, 22, 0.47 and 0.72, we could have given the final answer as 0.795m.

Exercise 1.2

A bedspring is 0.20 m high when not compressed. If the spring constant is 300Nm^{-1}, find the height of the spring when it is compressed by a load of 15N. Assume that Hooke's law is obeyed and remember that you need to subtract the 'extension' this time.

How we measure forces

It is not possible to measure a force directly because it is a rather abstract concept. Instead we determine the size of a force by measuring the effect which it has on another object or material. This usually involves springs again but it can involve other effects such as the bending of a metal strip.

A newtonmeter (or forcemeter as it is sometimes called) usually determines a force by measuring the extension or compression of a spring. As long as there is an end stop which prevents the spring from being overstretched, the extension of the spring will be proportional to the force applied. Pulling forces can be measured using springs which can be stretched and pushing forces by springs which can be compressed. One particular newtonmeter is able to measure forces up to ten newtons. If the spring inside it were replaced by one with a spring constant ten times as big it would then read up to 100 newtons.

Kitchen and bathroom scales are calibrated in units of mass e.g. grams, kilograms, pounds. In fact most of them have a spring inside which is compressed so they are measuring weight, not mass. This is not a problem on earth since gravity is fairly constant and they are calibrated on the assumption that a force of gravity of one newton means that the object has a mass of 100g. If we took our kitchen scales to the moon however, we would need six burgers to achieve a reading of 100g. This is not because the mass has changed, it is because the scales are really measuring weight which has changed because the force of gravity is less.

Another opportunity to feel a force of one newton is to pick up a Cox's apple. You get about ten apples to the kilogram so each apple must have a mass of 100g. Since the force of gravity is about ten newtons per kilogram on earth, the force on one apple must be about one newton.

It is time again for more practical exercise, perhaps you should pick up that apple with a force of one newton and then eat it.

What we should have learnt from this chapter

► Forces are pushes and pulls and are measured in newtons.

► On earth the force on a one kilogram object is about ten newtons.

► Hooke's law states that the extension of a spiral spring is proportional to the force applied.

► If a spring is overstretched it is said to have exceeded its elastic limit and Hooke's law is no longer obeyed.

► After the elastic limit the spring no longer returns to its original length when the force is removed.

► $F = kx$ where F is the force applied, x is the extension and k is the spring constant.

► Forces can be measured using forcemeters or newtonmeters.

Tutorial

Seminar discussion

Hookes law states: 'The extension of a spiral spring is proportional to the forced applied'. Discuss what happens when the elastic limit is reached.

Practical assignments

Investigate the effects of stretching on a spring. Find the spring constant and the elastic limit of the spring(s) you are investigating.

Study tips

It is always a worthwhile practice to read through some work in physics and then convince yourself first. By this we mean that you should ensure that you feel comfortable with your understanding of a concept.

The next stage is to convince another person. This process then confirms your own understanding or it illustrates a gap in your understanding.

2

Combining Forces

One-minute summary – Force, velocity and acceleration are vectors. This means that it is not good enough just to describe their size (or magnitude). The direction in which they act also has a large effect on the change which the force is able to cause. In this chapter we will:

▶ see that forces which act along the same line are easy to deal with
▶ learn that forces can be combined by using a parallelogram
▶ learn that forces can also be combined using a vector triangle or polygon
▶ learn how forces can be resolved into two directions at right angles.

Combining forces which act along the same line

If two people are pushing a car from the back, the combined effect can be found by adding the two forces together. A force of 40N and a force of 50N will give a combined force of 90N. It is similar to the fairy story about the enormous turnip. It takes the combined force from the husband, the wife... the cat and the mouse acting in the same direction to produce a resultant which is big enough to exceed the combined force of all the roots which are holding it in the soil.

If a car is being pushed from the back with a force of 70N and being pushed from the front in the opposite direction with a force of 60N, the **resultant** is 10N forward. In a tug of war competition, the members of a team are adding their forces together. The team which can produce the biggest combined force, so that the resultant in their direction is the winner.

Using a parallelogram

Figure 2(a) shows two ropes being used to pull a car along. Each rope is acting at an angle of 35° to the horizontal which is the straight forward direction. Figure 2(b) shows that the sides of a parallelogram can be used to represent the forces both in size and direction. The resultant force is then given in both size and direction by the diagonal of the parallelogram. The answer may be obtained by drawing the parallelogram to scale or by using the cosine rule on one side of the triangles found in the parallelogram.

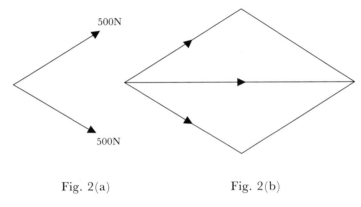

500N

500N

Fig. 2(a)   Fig. 2(b)

It can be seen from the diagram that this is a fairly straightforward example because the forces are equal and acting at equal angles to the forward direction. This means that the resultant will also be in the forward direction. See if you can find the size of the resultant yourself, either before or after reading the slightly more difficult example given below.

Example
The same car as above is now to be pulled by a force of 600N acting at 40° to the forward direction and a second force of 700N acting at 30°. (Figure 3(a).)

This can be solved either by scale drawing or by using the cosine rule and sine rule. Figure 3(b) shows the diagram for the scale drawing which is drawn to scale but due to printing methods the scale cannot be stated.

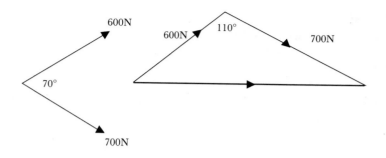

Fig. 3(a)   Fig. 3(b)

Using the cosine rule:

$$a^2 = b^2 + c^2 - 2bc\cos A$$
$$R^2 = 600^2 + 700^2 - 2 \times 600 \times 700 \times \cos 110°$$
$$= 360000 + 490000 - 840000 \times \cos 110°$$
$$= 850000 - (-287296.92)$$
$$= 850000 + 287296.92$$
$$= 1137296.9$$

therefore R = 1066

R = 1070N correct to three significant figures.

Using the sine rule to find the angle from the 600N force:

$$\frac{a}{\sin A} = \frac{b}{\sin B} = \frac{c}{\sin C}$$

$$\frac{700}{\sin A} = \frac{1066}{\sin 110}$$

$$\sin A = \frac{700 \times \sin 100}{1066}$$

$$= 0.6710$$

$$A = 38°$$

The angle between the resultant force and the forward direction is
$40° - 38° = 2°$ towards the 600N force.

Exercises

2.1 Find the size and direction of the resultant of a force of 15N and a force of 25N acting at an angle of 35°.

2.2 Find the size and direction of the resultant of a force of 40N and 60N acting at right angles.

2.3 A force of 35N and 60N are acting at an angle of 120°. Find the resultant.

Using a vector triangle

This method is very similar to the one described above except that it can be applied to more than three forces in which case it would produce a vector polygon. It is important when drawing the triangle that the directions of the forces are correct. If you are finding the equilibrant of two or more forces they will all point in the same direction round the polygon. If you are finding the resultant of two or more forces it will point the opposite way. This is because the **resultant** and the **equilibrant** are equal and opposite.

Example

Two wires are holding up an acrobat weighing 800N. One wire is at an angle of 60° to the vertical and the other is horizontal. Find the tension in each wire.

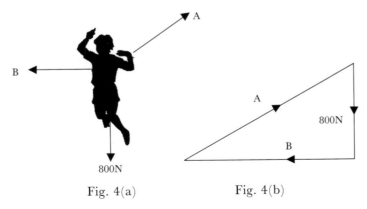

Fig. 4(a)   Fig. 4(b)

This is a fairly simple example since it involves two forces at right angles and will therefore produce a right-angled triangle. Figure 4(a) shows the forces on the acrobat and Figure 4(b) shows the vector triangle.

$$\text{sine} = \frac{\text{opposite}}{\text{hypotenuse}}$$

$$\sin 30 = \frac{800}{A}$$

$$A = \frac{800}{\sin 30}$$

$$= 1600\text{N}$$

$$\text{tangent} = \frac{\text{opposite}}{\text{adjacent}}$$

$$\tan 30 = \frac{800}{B}$$

$$B = \frac{800}{\tan 30}$$

$$= 1385$$

$$= 1390\text{N} \text{ correct to three significant figures.}$$

Example

We will now make the example more difficult by changing the horizontal wire to one which is at 45° to the vertical as shown in Figure 5(a). The triangle is now like the one shown in Figure 5(b) and we will need to use the sine rule.

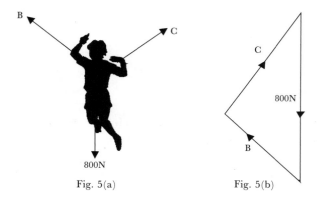

Fig. 5(a)   Fig. 5(b)

Using the sine rule:

$$\frac{a}{\sin A} = \frac{b}{\sin B} = \frac{c}{\sin C}$$

$$\frac{800}{\sin 85} = \frac{b}{\sin 45} = \frac{c}{\sin 60}$$

$$\frac{800}{\sin 85} = \frac{b}{\sin 45}$$

$$b = \frac{800 \times \sin 45}{\sin 85}$$

$$= 568N$$

$$\frac{800}{\sin 85} = \frac{c}{\sin 60}$$

$$c = \frac{800 \times \sin 60}{\sin 85}$$

$$= 695N$$

The wire at an angle of 60° to the vertical has a tension of 568N and the wire at 45° has a tension of 695N.

Exercise 2.4
A metal ball of mass 100Kg (weight = 1000N) is support by a chain which is initially vertical. Another horizontal chain is used to pull the ball to one side until the other chain makes an angle of 30.0° with the vertical. Find the tensions in each of the chains by constructing a vector triangle.

Resolving forces

A force can be considered as being the resultant of two forces acting at right angles to each other. Figure 6(a) shows a force of 27.0N acting at an angle of 35.0° to the horizontal. Figure 6(b) shows how to determine the horizontal and vertical components of the force.

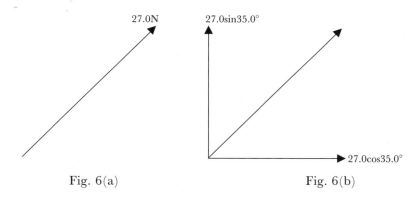

Fig. 6(a)   Fig. 6(b)

The horizontal component is 27cos35 = 22.1N. The vertical component is 27sin35 or 27cos(90 − 35) = 15.4N. Therefore the original force could be replaced by the two new forces. The two directions chosen do not have to be horizontal and vertical but they must be perpendicular to each other.

If a body is in equilibrium the total vertical components acting upwards must equal the total vertical components acting downwards. The horizontal components must match in the same way.

Example

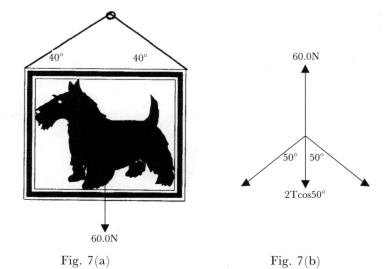

Fig. 7(a)   Fig. 7(b)

Figure 7(a) shows a picture hanging from a hook. Figure 7(b) shows the horizontal and vertical components of the forces in the string and on the hook. The picture is supported by the vertical reaction of the hook which opposes the downward force of the picture. This idea is discussed further in Chapter 9.

The horizontal components are both Tcos 40 and therefore cancel each other out. Considering the vertical forces:

$$2T\cos 50 \quad = \quad 60.0$$

$$T \quad = \quad \frac{60.0}{2 \cos 50}$$

$$= \quad 46.7N$$

The tension in each wire is 46.7N. The fact that it is the same wire hanging over a hook does not affect the vertical components. The tensions just behave as if there were two separate pieces of wire.

Example

In this example we will avoid any forces acting in convenient directions. A boat is tied up in a dock by three ropes acting at an angle to each other. Rope A is parallel to the length of the boat and is attached from the stern to the dock. Rope B and rope C are attached to the bow of the boat. Rope B makes an angle of $20°$ to the forward direction and rope C which has a tension of 500N makes an angle of $35°$ to the forward direction. In physics it is important to be able to draw a good diagram. Draw the diagram for this question. Label the direction parallel to the hull x and the direction perpendicular to the hull y. The correct diagram is given in the answers as Figure 25.

Resolving the forces along direction y:

$$B\cos 70 \quad = \quad 500\cos 55$$

$$B \quad = \quad \frac{500 \times \cos 55}{\cos 70}$$

$$= \quad 838.51N$$

$$= \quad 839N$$

Resolving the forces along direction x:

$$A = B\cos20 + 500\cos35$$
$$= 838.51\cos20 + 500\cos35$$
$$= 1197.52$$
$$= 1200N$$

Exercise 2.5

A wooden crate is suspended by two wires. One wire has a tension of 1000N and is at an angle of 35° to the vertical. The other wire has a tension of 1200N. Find, by resolving forces, the weight of the crate and the angle of the other wire.

What we should have learnt from this chapter

▶ A force is an example of a vector. When adding vectors we must consider their direction as well as their size.

▶ If the forces are acting along the same line they can just be added or subtracted as if they were scalar quantities.

▶ The resultant of two forces can be found by constructing a parallelogram of forces.

▶ A polygon of vectors is another way that can be used to find the effect of two or more forces.

▶ A force, like any other vector can be resolved into two directions at right angles.

Tutorial

Seminar discussion

Explain how to use a vector triangle.

Study tips

Revisit the sine rule from GCSE mathematics. Look at how you rearrange the equation to isolate the unknown variable.

3

Forces Which Change Velocity

One-minute summary – The velocity of a body is its rate of change of displacement with time (in a given direction). Speed is the rate of change of distance with time and does not take direction into account. A force can change both the speed and the direction of movement of a body. In this chapter we will learn:

▶ what is meant by inertia
▶ how forces make things accelerate and decelerate
▶ what a newton really is
▶ how a force can change the direction of a moving object
▶ about the forces which act on a body when its velocity changes very quickly.

What is meant by inertia?

Inertia is a measure of the resistance which an object has to a change in velocity. It has nothing to do with friction, although friction will usually either help or hinder the process of changing velocity. While thinking about inertia we are going to imagine an impossible situation where there is no friction. If we go back to our burger and try to change its speed we will need a certain force. If we pile three burgers on top of each other we will need a force three times as big to cause the same speed change. If there are five burgers the force is five times as big. We can see that as we increase the mass we increase the inertia. **Inertia is proportional to mass**.

A similar pattern is seen when slowing an object down. We are aware that an oil tanker takes several kilometres to stop, so the person in control must think ahead to avoid any hazards. Small ships must get out of the way. The tanker has a large mass and therefore a large inertia. Newton's first law of motion deals with inertia.

A body stays at rest or travelling with constant velocity unless acted upon by an external force.

Like all the laws in physics, each textbook will give slightly different wording for a law. The important fact is that they all mean the same thing. This law tells us that an object will not suddenly start to move from rest unless it is pushed and if already moving it will not change either speed or direction unless it is forced to do so. This is due to its inertia.

Making things accelerate

Whenever we change either the direction of travel or the speed of a moving body we are changing its velocity and therefore causing it to accelerate. If we are slowing it down we call it deceleration or a negative acceleration. Newton's second law of motion originally stated that the rate of change of momentum is proportional to the applied force. We find it more useful to state it in a completely different way.

The acceleration of a body is proportional to the applied force and inversely proportional to its mass.

The relationship is usually written as $F = ma$ where F is the applied force in newtons, m is the mass of the body in kilograms and a is the acceleration in ms^{-2}. This relationship is used to define the newton as a unit of force.

One newton is the force which will cause a body of mass one kilogram to have an acceleration of one metre per second per second (1ms^{-2}).

Example
Find the force needed to cause a body of mass 150kg to accelerate at 3.0ms^{-2}.

$$\text{From Newton's second law } F = ma$$
$$F = 150 \times 3.0$$
$$= 450\text{N}$$

Example

When a force of 525N is applied to a body, its speed changes from 10.0ms^{-1} to 20.0ms^{-1} in 2.50s. Find the mass of the body.

$$\text{acceleration} \quad = \quad \frac{\text{change in velocity}}{\text{time taken}}$$

$$= \quad \frac{20.0 - 10.0}{2.5}$$

$$= \quad 4.00\text{ms}^{-2}$$

From Newton's second law

$$F \quad = \quad ma$$

Therefore $m \quad = \dfrac{F}{a}$

$$= \frac{525}{400}$$

$$= 131\text{Kg (correct to three significant figures)}$$

Exercises

3.1 A car of mass 1000kg is towing a trailer of mass 750kg. Find the force needed if the acceleration of car and trailer is to be 1.5ms^{-2}. Assume there is no frictional force to overcome but the required friction between tyres and road is present.

3.2 A large delivery lorry is pulling a trailer. The lorry has a mass of 10000kg and the trailer has a mass of 5000kg. Find:
(a) the force needed to accelerate the lorry and trailer at 2.0ms^{-2}
(b) the force needed to accelerate just the trailer
(c) the tension in the coupling between the two.

What about a person in a lift?

This can have several different answers, depending on what is happening to the lift.

1. If the lift is stationary the effect felt by the occupants will be just the same as if they were standing on the ground. They will experience the normal force of gravity as their weight

pushes them against the floor of the lift which is supported by cables.

2. If the lift is moving either upwards or downwards at a steady speed, there is no accelerating force so the situation is exactly the same as if it were stationary.

3. If the lift is accelerating upwards, the occupants will experience an increased force between the floor and their feet. This is because the upwards force from the lift must not only balance their weight, but must also provide the upward accelerating force. The resultant acceleration will therefore be the acceleration which the force would produce under normal circumstances minus the acceleration due to gravity.

4. If the lift is accelerating downwards, the force between the floor and the person is less than normal, so they appear to be lighter than they would normally be. Gravity will therefore assist the acceleration.

What about free fall?

A person who is falling is effectively weightless. There is a schools' programme on gravity which was made several years ago that shows a boy jumping from a diving board holding a newtonmeter with a brick attached. This programme, which is still shown from video in many schools, demonstrates clearly that the reading on the newtonmeter falls to zero while the brick is falling. Clearly both the boy and the brick are weightless during free fall.

Later in the same programme people are shown being made to experience weightlessness in an aircraft. This is achieved by making the aircraft fly along a path which is the upper part of a vertical circle. If the radius of the circle and the velocity of the aircraft are carefully matched, its weight just matches the centripetal force required to keep it in the circle. This makes the plane and its contents appear weightless. (The theory of this is explained in Chapter 6.)

Forces which oppose

The most common opposing force is **friction**. When a force is applied to a body to move it over a surface, the opposing force of friction is initially equal and opposite to the applied force. This means that there is no resultant force and no acceleration occurs. Friction only opposes motion. If there is no applied force, there is no friction. As the applied force is increased, a point is reached when the frictional force can increase no more. This is when acceleration begins. The resultant force is then equal to the difference between the applied force and the frictional force.

Example
A box of mass 50kg is being pulled across a floor using an applied force of 100N. If the frictional force is 40N, find the acceleration of the box.

Resultant accelerating force on the box $= 100 - 40 = 60$N
From Newton's second law, $F = ma$

$$\text{Therefore } a = \frac{F}{m}$$

$$= \frac{60}{50}$$

$$= 1.2\text{ms}^{-2}$$

What causes friction?

Although a surface may look smooth, if magnified sufficiently it is possible to see bumps and depressions. The bumps from one surface get caught in the depressions in the other surface and this makes it more difficult for the surfaces to slide over each other. A wooden surface has a rougher surface than a polished metal surface so the friction is greater. This effect is shown in Figure 8.

Fig. 8

Applying a wax polish to each surface can reduce the friction between wooden surfaces. This fills up the depressions so that they do not trap the bumps from the other surface. A lubricant such as oil on a metal surface lifts the surfaces apart so that they move over each other on a cushion of oil.

Does friction occur in liquids and gases?

Liquids and gases are collectively known as **fluids**. Fluid friction is known as **viscosity**. Liquids such as golden syrup or oil are said to have a high viscosity. Liquids are able to flow because the layers of particles are able to flow over each other. In liquids with a high viscosity, the forces between the particles are stronger so they flow more slowly. This means that they flow through pipes more slowly and objects fall through them more slowly. The idea of forces between particles is dealt with again in Chapter 9.

What is terminal velocity?

When a body is falling through a fluid, it experiences an opposing force which is dependent on the viscosity of the fluid. This opposing force also increases with the velocity of the falling body. A falling body does so because of the force of gravity which acts on it. There is an opposing force caused by the surrounding fluid. As the body accelerates the opposing force increases. This decreases the resultant downward force which is causing the acceleration.

Eventually the upward force due to the fluid is equal to the force of gravity on the object (its weight) and the resultant downward force is zero as shown in Figure 9. The body has reached its terminal velocity and no further acceleration occurs. A ball bearing falling through oil reaches a terminal velocity after falling just a few centimetres. A free-fall parachutist falls many metres before the upward force is great enough to equal his or her weight. The terminal velocity is much too fast to land safely. When the parachute opens, the upward force is increased and there is a resultant upward force for a time until a new lower terminal velocity is reached which makes it safe to land.

Drag

Weight

Fig. 9

Can velocity change without changing speed?

Yes, it can. If, for example, a middle-distance athlete is running round a track with paces of constant length and a constant number of paces per second, he or she is moving at a constant speed. When running along the straight sections of the track the velocity is constant and no acceleration is occurring. At the ends of the track the direction of motion is changing so velocity is changing and acceleration is taking place. We will see in Chapter 6 that the acceleration is directed towards the centre of the circle of which the curve would form part.

A satellite in orbit round the Earth is travelling at constant speed but the velocity is changing. There is acceleration towards the centre of the Earth which keeps the satellite moving in a circular path.

What is impulse?

In order to understand impulse we need to understand a little about momentum. Momentum is the product of mass and velocity. In a collision between two bodies momentum is always conserved. At a fairly simple level, if two cars collide the following relationship will be true. Since momentum is a vector the direction needs to be taken into account.

Before collision		After collision	
car one	car two	car one	car two
mass × velocity +	mass × velocity =	mass × velocity +	mass × velocity

For a particular car e.g. car one the change of momentum is called **impulse**. Impulse is also equal to force × time. The force referred is the force which acts on a body during the momentum change. The time is the time it takes for the momentum change to occur. In order to reduce the force on passengers during a car crash, cars are designed to make the time for which the force acts as long as possible. Crumple zones are built into the car so that it does not change its velocity immediately but does so over a very short period of time. The softer surface beneath a swing in a play area makes the injuries less severe if a child were to fall on its head. A concrete surface would stop the child suddenly, resulting in a large force to the head. A surface which is slightly softer (but not too soft) provides a longer deceleration time and therefore a smaller deceleration force.

Example

A child of mass 30kg falls from a swing and hits the ground moving at $4.5ms^{-1}$. Find the force exerted on the child if it comes to rest in:

(a) one hundredth of a second and
(b) one second.

(a) The final momentum is zero so the change in momentum is
$30 \times 4.5 - 0$

$$\text{force} \times \text{time} = \text{change in momentum}$$
$$\text{force} = \frac{\text{change in momentum}}{\text{time}}$$
$$= \frac{30 \times 4.5 - 0}{0.01}$$
$$= 13500N$$

(b)

$$\text{force} = \frac{\text{change in momentum}}{\text{time}}$$

$$= \frac{30 \times 4.5 - 0}{1.0}$$

$$= 135\text{N}$$

Exercises

3.3 A woman of mass 65kg is travelling in a car which crashes into a wall and is brought to rest in 0.01s. If the car was travelling at 39ms^{-1}, find the force on the woman.

3.4 A bungee jumper reaches a velocity of 45ms^{-1} before being brought to rest gradually by the stretching elastic over a period of 5.1 seconds. If the mass of the jumper is 60kg, find the force which acts on her.

What we should have learnt from this chapter

▶ Inertia is proportional to mass and is a measure of a body's resistance to change in velocity.

▶ Newton's first law of motion states: a body stays at rest or travelling with constant velocity unless acted on by an external force.

▶ A force is able either to change the speed of an object and/or the direction in which it is travelling.

▶ Newton's second law of motion can be stated as: the acceleration of a body is proportional to the applied force and inversely proportional to its mass.

▶ This can be written as $F = ma$ where F is the applied force in newtons, m is the mass of the body and a is the acceleration in ms^{-2}.

▶ A force of one newton will give a body of mass one kilogram an acceleration of 1ms^{-2}.

▶ Friction is a force which opposes motion.

▶ Friction is caused by the irregularities in the surfaces.

▶ Friction in liquids is known as viscosity.

▶ A body reaches its terminal velocity when the opposing forces due to friction equal the accelerating forces. No further acceleration can then take place.

▶ The velocity of an object moving at constant speed will change if the direction of movement changes.

▶ The dangerous forces, which act on a person when they are made to stop, can be reduced if the stopping time is increased.

Tutorial

Seminar discussion

Explain what happens to inertia when mass is increased.

Study tips

Be clear in your own mind what is meant by a fluid. You should be able to define a liquid from memory and to describe what is meant by viscosity.

4

Forces Which Make Things Turn

One-minute summary – The distance between the point at which the force is applied and the point about which the object is turning controls the turning effect of a force. In this chapter we will learn:

- ▶ what is meant by the moment of a force
- ▶ how the principle of moments helps us to calculate the conditions needed for equilibrium
- ▶ how to deal with a pair of equal turning forces acting in the same direction such as those applied to turn on a tap
- ▶ the meaning of torque
- ▶ that for equilibrium both the resolved forces and the turning forces must be in equilibrium.

Moment of a force

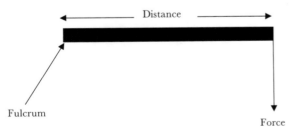

Fig. 10

The moment of a force is a measure of its turning effect. Increasing the distance from the place where the force is applied to the turning point increases the moment of a force. The turning point is usually called the **fulcrum**.

A simple definition of moment of a force could be:

moment = force × distance from the fulcrum

As the exercises become more sophisticated this will need to be:

moment = force × perpendicular distance from the point of application to the fulcrum

You can get an idea of how distance affects the moment of a force by finding a door which is unlatched and does not have an automatic closing device on it. Open the door by using your little finger to apply a force at the edge of the door which is furthest from the hinge. Now do the same thing but apply your force a few centimetres from the hinge side. The moment needed to open the door is the same in each case. In the first instance the force needed was much smaller because the distance from the fulcrum was much greater.

Example

A force of 10N is applied at the outer edge of a door in order to open it. The door is 0.70m wide. What force will need to be applied at 15cm from the hinge side?

In the first instance:

$$\text{moment} = \text{force} \times \text{distance from the fulcrum}$$
$$= 10 \times 0.70$$
$$= 7.0\text{Nm}$$

In the second instance, the moment will be the same:

$$\text{force} = \frac{\text{moment}}{\text{distance}}$$
$$= \frac{7.0}{0.15}$$
$$= 46.6666$$
$$= 47\text{N}$$

Principle of moments

The principle of moments states:

> **For a body in equilibrium, the sum of the anticlockwise moments is equal to the sum of the clockwise moments about the same point.**

The simplest application of this is with a seesaw. We must assume for the moment either that the seesaw is weightless or that it is balanced about its mid-point.

Example

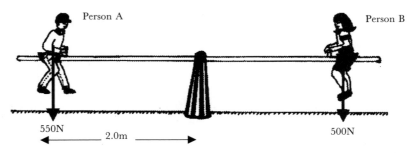

Fig. 11

Figure 11 shows a seesaw which has person A with a weight of 550N sitting 2.0m from the fulcrum. We need to find out where person B who weighs 500N must sit if the seesaw is to be in equilibrium.

Taking moments about the fulcrum:

$$550 \times 2.0 = 500 \times \text{distance B}$$

$$\text{distance B} = \frac{550 \times 2.0}{500}$$

$$= 2.2\text{m (to 2 sig. figs.)}$$

Example

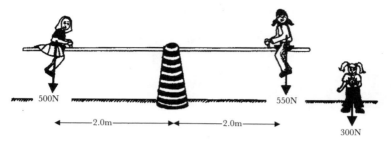

Fig. 12

Figure 12 shows a seesaw which has a person of weight 500N sitting 2.0m from the fulcrum. A second person of weight 550N is sitting 2.0m from the fulcrum on the other side. Where must a child of weight 300N sit if the seesaw is to balance?

Taking moments about the fulcrum:

$$500 \times 2.0 + 300 \times d = 550 \times 2.0$$

$$1000 + 300d = 1100$$

$$300d = 100$$

$$d = 0.33\text{m from the fulcrum}$$

Example

A mobile to hang over a child's cot consists of a very light rod of negligible weight of length 0.4m. A fluffy rabbit of mass 56.0g is attached to one end and a fluffy dog of mass 78.3g is attached to the other. Find the distance from the fluffy rabbit to the suspending thread if the mobile is to hang level.

Let the distance from the fluffy rabbit be y.

This means that the distance from the fluffy dog is $0.4 - y$.

Taking moments about the suspending thread:

$$\begin{array}{ccc} \text{moment of the rabbit} & = & \text{moment of the dog} \\ \text{about the thread} & & \text{about the thread} \end{array}$$

$$0.056 \times 10 \text{ x } y = 0.0783 \times 10 \times (0.4 - y)$$

$$0.56y = 0.3132 - 0.783y$$

$$0.56y + 0.783y = 0.3132$$

$$1.343y = 0.3132$$

$$y = \frac{0.3132}{1.343}$$

$$y = 0.23\text{m}$$

Exercises

4.1 A tightrope walker is using a pole 4m long to help his balance. He is walking across a waterfall and is carrying his lunch to have on the other side. On one end of the pole there is a bottle of drink of mass 2.0kg and on the other end there is his lunch box of mass 1.5kg. Where must he support the pole?

4.2 The passengers are using a plank of wood to lift one corner of a car so that the wheel can be changed by the driver. The upward force necessary on the car is 2500N. The plank has its end resting on the ground under the car and it makes contact with the car 0.5m from that end. If the length of the plank is 3m find the upward force which must be exerted at the end if the car is to be lifted. You may ignore the weight of the plank.

Centre of gravity

The **centre of gravity** of a body is the point about which the whole weight of the body may be considered to act. With a simple shaped object such as a ruler the centre of gravity is the balance point and is in the middle of the object. With a hollow ball of even thickness and density the centre of gravity is in the middle of the space at the centre of the ball. For a cone shape, the weight is concentrated more towards the base so the centre of gravity is not in the middle. It can be shown mathematically that it is one third of the way up from the base. For a plain wedding ring the centre of gravity will be at the centre of the hole in the middle. An engagement ring is not symmetrical so the centre of gravity will be off centre due to the weight of the diamond. Up to now we have only considered questions about planks which had no weight of their own or were supported at their centre of gravity. This meant that the weight of the plank acted through the support and therefore did not affect the balance.

If the object is not supported at its centre of gravity, an additional force equal to its weight will act at the centre of gravity. This is shown in the following example.

Example

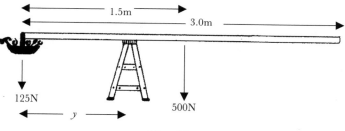

Fig. 13

Figure 13 shows a plank of wood which has a weight of 500N and which is balanced by a tool bag which is hung on the end. If the plank is 3.0m long and the bag has a weight of 125N, find the distance between the bag and the trestle.

Let the distance between the bag and the trestle be y

Taking moments about the trestle

$$125y = 500 \ (1.5 - y)$$

$$125y = 750 - 500y$$

$$625y = 750$$

$$y = \frac{750}{625}$$

$$y = 1.2m$$

Therefore the distance from the bag to the trestle is 1.2m.

Exercise 4.3

A man is carrying a ladder which is not of uniform weight, so its centre of gravity is not in the middle. The ladder is 4m long and has a weight of 300N. There is a bag of tools of weight 135N hanging on the light end. The man is supporting the ladder at a distance of 1.8m from the tool bag. What is the distance between the tool bag and the centre of gravity?

Reactions at supports

If an object, such as a plank, is supported from below (such as a decorator might have a plank across two trestles) each trestle must be exerting an upward force equal and opposite to the downward force acting on it. This upward force is known as the reaction at the support. If the plank is to be in equilibrium, the total upward forces must equal the total downward forces.

Example

A uniform plank weighing 100N and of length 5.0m is supported by a trestle 0.5m from each end. The decorator of weight 700N is standing 1.5 m from one end and a can of paint of weight 100N is standing 2.0m from the other end. Find the reaction at each support.

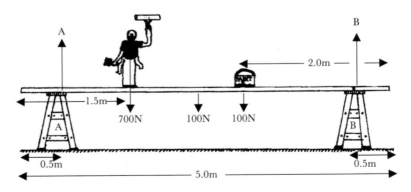

Fig. 14

We will first take moments about the trestle which is marked A on Figure 14, so the moment of A will be zero. We will then take moments about the other trestle so that its moment will be zero.

Taking moments about A:

$$(700 \times 1.0) + (100 \times 2.0) + (100 \times 2.5) = B \times 4.0$$

$$700 + 200 + 250 = B \times 4.0$$

$$1150 = B \times 4.0$$

$$B = \frac{1150}{4.0}$$

$$B = 287.5N$$

$$B = 288N \text{ (3 sig. fig.)}$$

Taking moments about B:

$$(100 \times 1.5) + (100 \times 2.0) + (700 \times 3.0) = A \times 4.0$$

$$150 + 200 + 2100 = A \times 4.0$$

$$2450 = A \times 4.0$$

$$A = \frac{2450}{4.0}$$

$$A = 612.5N$$

$$A = 613N \text{ (3 sig. fig.)}$$

We can check that the total upward forces equal the total downward forces:

$$\text{Upward forces} = \text{Downward forces}$$

$$287.5 + 612.5 = 100 + 100 + 700$$

$$900N = 900N$$

Exercise 4.4

The army has built a temporary bridge across a river. The weight of the main part of the bridge is taken by two supports placed 12.5m apart. A lorry of mass 3000kg has left support A proceeding towards support B but has broken down so that its centre of gravity is 4.5m from A. If the deck of the main part of the bridge has a weight of 65000N, what is the reaction at each support?

What happens when the force is not perpendicular?

To calculate a moment the force and the distance measured must be perpendicular. This can be achieved in one of two ways. You can either take the component of the force in a direction perpendicular to the distance or you can measure the distance in a different direction so that it is perpendicular to the force. The example below takes the component of the force.

Example
A sign for a village pub in Little Middlethrop consists of a hanging board suspended from a horizontal rod and weighing 450N. The rod is 0.90m long and a metal rod which makes an angle of 40° with the wall is attached to the far end of the rod and takes part of the weight. The rod is attached to the wall at the other end. In order to make it stand out a bit, the board is not mounted centrally on the rod. Its centre of gravity is 0.60m from the wall. Find the tension in the rod and the direction and size of the reaction at the wall.

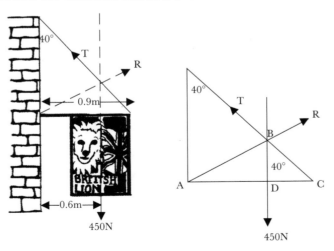

Fig. 15

As shown in the Figure 15, the line through the centre of gravity of the sign, the tension and the reaction must all pass through the same point. This allows us to find the direction of the reaction at the wall.

For the triangle BCD, $\tan 40 = \dfrac{CD}{BD}$

$$BD = \frac{CD}{\tan 40} = \frac{0.30}{\tan 40} = 0.357526\text{m}$$

$$\tan BAD = \frac{BD}{AD} = \frac{0.357526}{0.60}$$

$$BAD = 30.789733°$$

Taking moments about A, (this will mean that the moment of R is zero):

$$T \cos 40 \times 0.90 = 450 \times 0.60$$

$$T = \frac{450 \times 0.60}{\cos 40 \times 0.90} = 391.62219 = 390\text{N}$$

Taking moments about C, (this will make the moment of T zero):

$$R \cos 59.210267 \times 0.90 = 450 \times 0.30$$

$$R = \frac{450 \times 0.3}{\cos 59.210267 \times 0.9} = 293.03231 = 290\text{N}$$

Check:

Total upward forces should equal the total downward forces

$$R \cos 59.210267 \times 0.9 + T \cos 40$$

$$= 293.03231 \times \cos 59.210267 + 391.62219 \times \cos 40$$

$$= 150 + 300$$

$$= 450\text{N} = \text{the downward force}$$

Exercise 4.5

A garden table is supported by four legs. When viewed from the side the legs are perpendicular to the table. When viewed from the end they each make an angle of 25° with the vertical and positioned 0.60m apart. If the mass of the table top is 20kg, find the reaction in each leg.

Couples and torques

A **couple** is a pair of equal forces which are acting on an object at equal distances from the fulcrum and tending to turn the object in the same direction. Two common examples of this are when we turn a tap on or off or when a driver turns a steering wheel. A **torque** is the moment of a couple and is equal to the value of one of its forces multiplied by the distance between them. This is the same as taking the moment of one of the forces and then doubling it.

Keeping things in equilibrium

When a body is being affected by a number of forces, in order for it to be in equilibrium, there must be no resultant linear force in any direction nor any resulting turning force.

What we should have learnt from this chapter

▶ The moment of a force on a body is a measure of the turning effect which it causes. Moment = force × perpendicular distance from the point of application to the fulcrum.

▶ The principle of moments states that for a body in equilibrium, the sum of the anticlockwise moments is equal to the sum of the clockwise moments about the same point.

▶ The centre of gravity of a body is the point about which the whole weight of the body may be considered to act.

▶ A couple is a pair of equal forces acting at equal distances on the opposite sides of the fulcrum. They produce turning moments in the same direction.

▶ A torque is the moment of a couple.

▶ A body can only be in equilibrium when the linear forces and the turning forces are balanced.

Tutorial

Seminar discussion
Explain what is meant by:

▶ The moment of a force.
▶ Torque.

Study tips
Create a PowerPoint slide show that contains definitions and diagrams that you can refer back to later for revision purposes.

5

Forces Which Do Work

One-minute summary – Work is done when a force moves its point of application. In this chapter we will:

► learn what is meant by work and its connection with energy
► understand when work is and is not being done
► find out how simple machines can reduce the force needed for a job
► find out what is meant by mechanical advantage.

Getting tired without doing work

A force does work only when it moves its point of application. Imagine someone trying to push a car along. If the driver has left the car in gear or has left the brake firmly on it will not be possible to move it. The person trying to push it may get very tired and finish up with aching muscles but if the car has not moved, no work has been done on it.

Measuring work

As mentioned already work involves both a force and movement. The definition of work done is:

work done = force × distance moved in the direction of the force

If the force is measured in newtons and the distance is in metres, the unit of **work** is the **joule** (J). (James Prescott Joule was the son of a Manchester brewer and did work on the heat produced when work is done. You may like to find out more about him.)

Example

A woman pushes a car a distance of 3m. If it requires a force of 100N, how much work does she do?

$$\text{Work done} = \text{force} \times \text{distance}$$
$$= 100 \times 3$$
$$= 300J$$

Example

A forklift truck lifts a packing case from the floor onto the top of a pile which is 6.0m high. If the packing case has a mass of 600Kg and the force of gravity is 10N Kg^{-1}, find the work done on the packing case.

The force required to lift the packing case will be equal and opposite to the weight.

$$\text{Weight} = \text{mass} \times 10$$
$$= 600 \times 10$$
$$= 6000N$$
$$\text{Work done} = \text{force} \times \text{distance moved}$$
$$= 6000 \times 6.0$$
$$= 36000J$$
$$= 36KJ$$

Work and energy

Energy is the capacity to do work. We get our energy from food. Machines get their energy from electricity or from fuels. Energy is needed in order to do work. Energy is also measured in joules. We need 100J of energy if we are to do 100J of work. This assumes that all the energy that is supplied gets to the place where it is required. This is not usually the case. We say that the system is not 100% efficient. The efficiency of a system compares the energy put into the system with the useful work done by the system.

$$\text{efficiency} = \frac{\text{useful work done by system}}{\text{total energy supplied to system}}$$

The result is often multiplied by 100 to give the efficiency of the system as a percentage. Thus an efficiency of 0.5 or 50% means that half of the energy is used as intended and the other half is dissipated in some other form, usually as heat.

Is the work done quickly or slowly?

The forklift truck used in the example above may do the work either quickly or slowly. The rate at which a system does work is called its power.

$$\text{Power} = \frac{\text{work done}}{\text{time taken}}$$

or

power is the rate of doing work

Example

Fork lift truck (a) can lift the packing case in 5.0 seconds but fork lift truck (b) takes 15.0 seconds. Find the power of each truck.

(a)

$$\text{Power} = \frac{\text{work done}}{\text{time taken}}$$

$$= \frac{36000}{5.0}$$

$$= 7200\text{W}$$

(b)

$$\text{Power} = \frac{\text{work done}}{\text{time taken}}$$

$$= \frac{36000}{15}$$

$$= 2400\text{W}$$

Exercise 5.1

An electric stair lift is able to take an old lady of mass 58kg up a flight of 20 stairs in 12 seconds. If each step is 25cm high, calculate the power of the lift.

Carthorse vs racehorse

Power can be demonstrated in different ways. A carthorse is able to pull a very heavy load because it has very strong muscles and can therefore exert large forces. A racehorse, however, is bred for speed. It would not be able to pull the heavy loads which a carthorse is able to pull. In our formula for power the carthorse is able to make the answer greater by making the force bigger. The racehorse cannot make the force bigger but it can make the time smaller. Both of these changes will increase the power.

Using levers as force multipliers

When we need to remove the lid from a can of paint we get a screwdriver, put the blade under the rim and use it as a lever to allow us to exert a larger force on the lid than we could normally. This is an example of a **force multiplier**.

A force multiplier allows us to use a small force in order to exert a larger force.

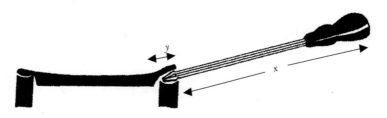

Fig. 16

Figure 16 shows how this type of lever works. This is a moments problem. The fulcrum is the rim of the paint can. Since the distance y is much smaller than the distance x, the force on the lid will be much greater than the force applied to the handle. This must be true so that the moments of the two forces are equal.

Is there a catch here since energy must be conserved? Notice that the distance which the handle moves down is much greater than the distance which the lid moves up. This means that the work done on the lid is just the same as the work done on the handle. Therefore energy is conserved. Some texts refer to this as a force/distance trade off.

Example

If the distance x from the lid to the fulcrum is 0.75cm and the distance from the fulcrum to the handle is 20cm, find the force exerted on the rim when the force applied to the handle is 20N.

To keep everything tidy we will work in metres, although it is not essential in this particular case.

$$0.75\text{cm} = 0.0075\text{m and 20cm is } 0.20\text{m}$$

$$\text{Moment of force on handle} = \text{moment of force on lid}$$

$$\text{Force x distance on handle} = \text{force} \times \text{distance on lid}$$

$$20 \times 0.20 = \text{force on lid} \times 0.0075$$

$$\text{Force on lid} = \frac{20 \times 0.20}{0.007}$$

$$= 530\text{N (2 sig. fig.)}$$

Exercises

5.2 A pair of long handled pruning shears has handles which are 80cm long. The distance from the pivot to the place when the branch is gripped is 6.5cm. If the handles are pushed together with a force of 50N, what force is applied to the branch?

5.3 There is a story from Greek legend about a soldier who had a son but while the boy was still a baby his father had to go to war. In case he was killed, the father wished to leave his son a sword to use when the boy had reached an age when it was safe to do so. He left the sword under a very heavy stone because he believed that he needed either to be strong enough to lift the stone or clever enough to move it without lifting. Of course the boy was both but he could have used a length of wood as a lever to act as a force multiplier. If the weight of the stone was 10,000N and the wood was 6.0m long, where would he need to place his small rock to act as a fulcrum so that the force needed was 500N?

Using levers to get more movement

When an angler casts a line, the handle of the rod is held still and becomes the fulcrum. The other hand is placed on the rod a small distance from the fulcrum and the small movement of this hand causes the top of the rod to move through a large distance. This could be described as a distance multiplier since a small movement of the hand causes a much larger movement of the tip of the rod which casts the line further out into the river.

Getting a mechanical advantage

When a simple machine such as a lever or a pulley system allows a small force to be increased to a larger force it is sometimes called getting a mechanical advantage. A force multiplier always gives a mechanical advantage greater than one. A mechanical advantage of less than one would mean that the force was smaller.

$$\text{mechanical advantage} = \frac{\text{load}}{\text{effort}}$$

What we should have learnt from this chapter

▶ Work done = force × distance moved in the direction of the force.

▶ Energy is the capacity for doing work.

▶ Work and energy are both measured in joules.

▶ Work needs force and movement; the application of a force without movement produces no work.

▶ Power is the rate of doing work.

▶ Levers can be used to reduce the size of the force which is needed to do a job. The lever is then acting as a force multiplier.

▶ If the force is made smaller the distance moved increases so energy is conserved.

▶ A force multiplier is said to produce a mechanical advantage.

Tutorial

Seminar discussion

1. Explain what a *force/distance trade off* means.
2. Who was it who suggested that given a place to stand they could move the Earth and what did they mean by this comment?

Study tips

Try modelling situations with levers and seeing what happens when the forces are applied. Then scale these up by drawing and calculating.

6

Forces Which Keep Things Moving in a Circle

One-minute summary – As its name suggests, this chapter will show how a body is kept moving in a circle when, according to Newton's first law of motion, it should carry on in a straight line. We will:

▶ sort out misunderstandings about centrifugal and centripetal forces

▶ learn how to calculate the size of the force needed to keep a body moving in a circle of given radius.

Centripetal and centrifugal

Many students are confused about the forces involved when a body is moving in a circle. To understand the situation we must start from Newton's first law of motion which tells us that the natural motion of a body is for it to continue at a steady speed in a straight line. This means that at any point in its motion, the natural tendency is for the body to move off along a tangent to the circle at that point. If it is to continue to move in a circle a steady force towards the centre is necessary to prevent it from 'going off at a tangent'. The force towards the centre is called the **centripetal force**.

The bodies therefore experience this force pulling them towards the centre and it appears to be working against an imaginary force pulling outwards. It is this non-existent force which has been given the name **centrifugal**.

Something to try

Choose an open space and tie a 'safe' object such as the cork from a

55

wine bottle to a piece of string. Whirl the cork round in a circle above your head and let it go suddenly at a known position. Note whether it goes out along a radius or 'goes off at a tangent'.

A force towards the centre

Now that we have established that a centripetal force is needed to keep an object moving in a circle, we can look at the size of this force.

$$F = \frac{mv^2}{r}$$

Where m is the mass of the body, v is the linear velocity and r is the radius of the circle. This same equation can also be expressed as

$$F = mr\omega^2$$

Where ω is the angular velocity in radians per second. 2π radians make one full circle (meaning that 2π radians equals $360°$).

We can look at the first equation in a number of ways. If the body is heavier i.e. has a larger mass it will:

▶ need a larger force to keep it travelling at the same speed and distance from the centre, or
▶ it will move to an orbit with a larger radius, or
▶ it will slow down.

Something else to try

This again uses the cork tied on a piece of string. You need to make the equipment a little more sophisticated, however. You will need something to act as a handle which will allow the string to pass freely along the middle of it (such as a piece of metal or plastic pipe). Alternatively, it could simply be a plastic bottle with the bottom cut out of it. If it is a bottle, place the stopper end at the top, nearest to the cork. You need to tie some kind of weight on to the other end of the string.

Hold the 'handle' and try to swing the cork above your head as slowly as you can while keeping the string to the cork nearly horizontal. Try to take note of the speed at which the cork is going round, i.e. the number of revolutions per minute. If you now speed up the number of revolutions per minute you should find that the cork moves outwards to a bigger orbit.

This means that v in the formula will have increased but the weight of the cork and the tension in the string have stayed the same, causing the radius to increase. You could bring the orbit back to the original size by increasing F which means putting a larger weight on the end of the string.

Example
Find the tension in the string when a stone of mass 500g is swung in a horizontal circle of radius 0.50m at a speed of 3.0ms^{-1}.

$$F = \frac{mv^2}{r}$$

$$= \frac{0.500 \times 3.0^2}{0.50}$$

$$= 9.0N$$

Example
A stone of mass 3.0kg on a piece of string is being whirled in a vertical circle of radius 19cm. Find the minimum speed at which the stone must move if it is to stay in a circular path.

To stay in the circular path, the tension in the string must be equal to or greater than the weight of the stone otherwise it will fall when it is at the top of the circle. (Note the connection between this problem and the aircraft flying in a vertical circle to simulate weightlessness.)

$$F = \frac{mv^2}{r}$$

$$v^2 = \frac{Fr}{m}$$

$$v^2 = \frac{3.0 \times 10 \times 19}{3.0}$$

$$v^2 = 190$$

$$v = 14ms^{-1}$$

Exercises

6.1 A car is travelling over a hump back bridge. The radius of curvature of the hump is 16.0m. If the car has a mass of 750kg, what is the minimum speed at which the car will lose contact with the bridge as it passes over it? If the mass of the car is doubled, how will this affect the answer?

6.2 A stone of mass 0.55kg on the end of a piece of string is spun in a vertical circle. The angular velocity of the stone is 15.3 radians per second and the radius of the circle is 0.47m. Find the maximum and minimum tension in the string.

6.3 When a pilot loops the loop in an aircraft the force with which they are pushed into their seat is found to vary at different points of the loop. Find the maximum and minimum force and state where they occur if the pilot has a mass of 75kg and is travelling in a circle of radius 520m at a speed of 70ms^{-1}.

6.4 A bucket of water is being swung in a vertical circle of radius 2.5m. What is the minimum speed in revolutions per second if no water is to be spilt?

Going round the bend

When a car or a bicycle is going round a bend in the road, a **force towards the centre** of the curve is necessary in order to get the vehicle round the corner. At slow speeds and fairly gentle bends the frictional force between the tyres and the road provides this force. This is fine so long as the force required does not exceed the maximum frictional force which can exist between the two surfaces. If this maximum force is exceeded the vehicle will skid instead of turning into the curve and an accident will result. A train is able to go round a gentle bend at relatively slow speeds because the required centripetal force is provided by the normal reaction between the outer wheels and the rail. This deals with slow speeds and gentle bends but cars, trains and bicycles usually need to go round tighter curves and at higher speeds. This can be achieved by the cyclist (both motor and other) leaning towards the centre of the curve. For cars and trains, a similar effect can be achieved by banking the track, i.e. making it slope towards the inside of the track. The effect of both of these is to produce a force towards the centre of the curve. In recent years there has been talk of tilting trains which would provide this force without having to bank the track and the degree of tilt would vary with the speed of the train.

How banking and leaning work

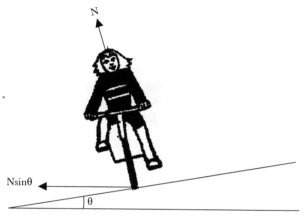

Fig. 17

Figure 17 shows how the centripetal force is provided by the horizontal component of the normal reaction of the surface. We need to be able to calculate the angle of banking which is required in a certain situation. The normal reaction is shown as N and the track is banked at an angle θ. The horizontal component of N is $N\sin\theta$ and this must be equal to the centripetal force. The vertical component of N is $N\cos\theta$ and this must equal the weight of the vehicle and contents or bicycle and rider.

$$N\sin\theta = \frac{mv^2}{r}$$
$$N\cos\theta = mg$$

If we divide these two expressions $\dfrac{N\sin\theta}{N\cos\theta} = \dfrac{mv^2}{mgr}$

$$\text{Tan } \theta = \frac{v^2}{gr}$$

It is interesting and useful to note that the required angle does not depend on the mass of the vehicle, therefore the angle is the same for a small car as for a juggernaut. It does, however, depend on the velocity. This means that on a normal road the angle is calculated to correspond to the safe driving speed on that stretch of road. In a cycling stadium the angle of the track increases so the rider can choose a path which matches the speed. On Formula One tracks the surface is banked only where safety demands it. Normally it is part of the skill required by the driver to change speed in an appropriate way to get round the corner as quickly as possible but without crashing. An aircraft is able to turn by dropping one of its wings and raising the other so that the horizontal component of lift can be used. (See Chapter 11.)

Example
At what angle must a road be banked if a car travelling at 10.5ms^{-1} is to get safely round the curve of radius 95m? A diagram similar to Figure 17 could be drawn to represent the car. In this case the weight and therefore the normal reaction would be spread over four wheels rather than two but it can be shown concentrated at one point.

Let θ be the required angle of banking.

$$\frac{N\sin\theta}{r} = mv^2$$

$$N\cos\theta = mg$$

$$\frac{\text{Tan } \theta}{gr} = v^2$$

$$= \frac{10.5^2}{10 \times 95}$$

$$= 0.1160526$$

$$\theta = 6.6°$$

Exercises

6.5 The track of a bobsled run is banked at an angle of 23.0° on one particular curve. If the radius of the curve is 165m, what is the optimum speed at which the curve should be taken?

6.6 A motor cyclist is going round a corner of radius 100m on a flat road at a speed of 13.8ms^{-1}. Find the angle to which they must lean if they are to get round the corner safely.

What we should have learnt from this chapter

▶ Centrifugal force is a misunderstanding of the tendency of a body to carry on in a straight line.

▶ Centripetal force is the force at right angles to the direction of travel which will keep a body moving in a circle.

▶ If the centripetal force is cut off the body will stop moving in a circle and just carry on in its current direction in a straight line.

▶ The size of the centripetal force is given by $F = \frac{mv^2}{r}$ or $F = mr\omega$.

▶ Increasing either the mass or the speed will mean that a larger force is needed to maintain a circular path of the same radius.

▶ When a vehicle is going round a bend, it is necessary to have a force towards the centre of the curve, i.e. a centripetal force.

▶ This centripetal force can be provided either by banking the road, so that the horizontal component of the normal reaction to the surface provides this force, or by the vehicle tilting towards the centre of the curve.

Tutorial

Seminar discussion
Explain what is meant by centripetal forces.

Study tips
Try video recording the practicals in this chapter where you rotate safe objects on a string. Look back at the video and make sure you can see how the centripetal force works.

7

Forces Caused By Fields

One-minute summary – Forces caused by fields are the most mysterious of all. We can feel the fridge door being closed by the force between the magnetic strip and the metal of the case but we can see nothing. Equally strange forces exist in gravitational fields and in electric fields. In this chapter we will learn:

▶ how these forces behave
▶ how their behaviour can be represented by formulae
▶ how they cause some of the greatest and some of the smallest phenomena in the universe.

Electric forces

Electric forces are experienced by charged objects in **electric fields**. Inversely we could define an electric field as:

A region where a body experiences a force due to the electric charge that it carries.

The direction of the field is the direction in which a positively charged body would experience the force. Figure 18 shows two electric fields. A pair of parallel plates causes each one. By convention, the direction of the field is always towards the more negative plate.

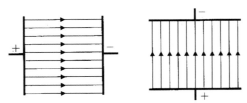

Fig. 18

Electric field strength is measured in **newtons per coulomb**. If the electric field strength is $10NC^{-1}$ it means that a body carrying a charge of one coulomb would experience a force of 10 newtons, whereas a body carrying a charge of two coulombs would experience a force of 20 newtons. It is impossible to measure the force on a charged body directly, especially if the body is very small. It is also difficult to measure the charge on a body with any accuracy. It can be shown, however, that for a uniform electric field of the type which you get between two parallel plates, the field strength is also given by the potential difference between the plates divided by the distance between them. Remember that if you are asked to define electric field strength you should give the definition which involves newtons per coulomb. The other relationship is just convenient for calculations and practical work.

$$\text{electric field strength} = \frac{\text{force}}{\text{charge}} \qquad \text{electric field strength} = \frac{\text{PD}}{\text{distance}}$$

Example

A potential difference of 5000V is maintained between two parallel plates which are 25mm apart. Find the electric field strength and also the force which would be exerted on a small body carrying a charge of 0.005C.

$$\text{electric field strength} = \frac{\text{PD}}{\text{distance}}$$

$$= \frac{5000}{0.025}$$

$$= 200000Vm^{-1}$$

$$= 200000NC^{-1}$$

$$\text{electric field strength} = \frac{\text{force}}{\text{charge}}$$

$$\text{force} = \text{charge} \times \text{electric field strength}$$

$$= 0.005 \times 200000$$

$$= 1000N$$

Example

A small drop of oil is held stationary in an electric field by a potential difference of 1000V between parallel plates with a separation of 10cm. If the charge on the drop is 0.00001C, what is the mass of the drop?

$$\text{electric field strength} \quad = \frac{\text{force}}{\text{charge}} \quad = \frac{\text{PD}}{\text{distance}}$$

$$\text{force} = \frac{\text{PD}}{\text{distance}} \times \text{charge}$$

$$= \frac{1000 \times 0.00001}{0.1}$$

$$= 0.01\text{N}$$

$$\text{mass of drop} = \frac{0.01}{10} \quad \text{(the force on 1kg is 10N)}$$

$$\text{mass of drop} = 0.001\text{kg} = 1\text{g}$$

Example

Examination boards like to set questions which involve more than one subject area. The example below is one of this type.

A small charged ball is hung from a stand by an insulated thread between two vertical plates with 500V between them and separated by 0.075m. The thread hangs at an angle of 4° to the vertical and the tension in the thread is 4.5×10^{-4}N. Find the electric force on the ball and hence its charge.

The electric force is horizontal and this is balance by the horizontal component of the tension in the thread.

$$\text{electric force} = 4.5 \times 10^{-4} \sin 4°$$

$$= 3.1390 \times 10^{-5}$$

$$= 3.1 \times 10^{-5}\text{N}$$

$$\text{electric field strength} \quad = \frac{\text{force}}{\text{charge}} \quad = \frac{\text{PD}}{\text{distance}}$$

$$\text{charge} = \frac{\text{force} \times \text{distance}}{\text{PD}}$$

$$= \frac{3.1390 \times 10^{-5} \times 0.075}{500}$$

$$= 4.7 \times 10^{-9}\text{C}$$

Exercise 7.1

An electric field is maintained between two vertical plates separated by a distance of 25mm. The PD between the plates is 1000V. Calculate the work done in moving a charge of $1.0 \times 10^{-14}\text{C}$ from one plate to the other.

Radial electric fields

The above examples concern charged bodies in uniform fields. We must also consider **radial fields**. Coulomb's law concerns the force between two charged bodies.

$$F = \frac{1}{4\pi\varepsilon_0} \frac{Q_1}{r^2} Q_2$$

Coulomb simply stated that the force between two charged bodies is proportional to each of the charges and inversely proportional to the distance between them.

ε_0 is a universal constant known as the permittivity of free space, Q_1 and Q_2 are the charges on the two bodies in coulombs and r is the distance between them in metres. This formula is used to calculate the forces between two charged particles. If the charges are the same, the forces repel but if they are different, they attract.

Example

Two alpha particles are separated by a distance of 5.0×10^{-7}m. The charge on each alpha particle is 3.2×10^{-19}C and ε_0 is 8.85×10^{-12} $\text{C}^2\text{N}^{-1}\text{m}^{-2}$. Find the force between the two particles. (An alpha particle is the same as a helium nucleus.)

$$F = \frac{1}{4\pi\varepsilon_0} \frac{Q_1 Q_2}{r^2}$$

$$\frac{1}{4\pi \times 8.85 \times 10^{-12}} \qquad \frac{3.2 \times 10^{-19} \times 3.2 \times 10 = 19}{(5.0 \times 1^{-7})^2}$$

$$= 3.7 \times 10^{-15} \text{N}$$

The electric field strength at a distance r from the centre of a charged body in a radial electric field is given by:

$$F = \frac{1}{4\pi\varepsilon_0} \frac{Q}{r^2}$$

Q is the charge on the body which is causing the field and the formula includes the permittivity of free space as before.

Exercise 7.2
In a hydrogen atom, the distance between the proton and electron is 0.50×10^{-10} m. Find the force of attraction due to their charges. ($e = 1.6 \times 10^{-19}$C and ε_0 is 8.9×10^{-12}Fm^{-1}.) (Note that we are ignoring the other forces which are acting in this situation.)

Gravitational forces

Gravitational fields behave in a very similar way to electric fields. The formulae have a similar structure to those in electric fields except charge is replaced by mass and $\frac{1}{4\pi\varepsilon_0}$ is replaced by G known as the universal gravitational constant. G should not be confused with g which is not a universal constant since it varies with the mass of the planet and the distance from the centre. The uniform gravitational field is not as important as the uniform electric field. The only near example which we have is that the Earth's gravitational field is taken to be approximately constant for the first 100km above the surface but this is only a convenient approximation.

Once again it is Sir Isaac Newton who is associated with this field of physics, as he is with so many others. Coulomb's law for the force between two charges is replaced by Newton's law of gravitation. This is:

$$F = G \frac{m_1 m_2}{r^2}$$

F is the force between two bodies of mass m_1 and m_2 separated by a distance r between their centres and G is the universal gravitational constant. This means that the gravitational forces between them attract all objects to each other but since the mass is so small these forces are undetectable. If one of the objects is a planet, the forces are much bigger because of the large mass involved. Notice that there is no repulsion between objects, only gravitational attraction.

Example

If you are sitting next to another person there is a gravitational attraction between you. Let us flatter ourselves and say that you and the other person have a mass of 50kg and the distance between your centres of gravity is 0.75m. Find the attractive force between you.

$$F = G \frac{m_1 m_2}{r^2}$$

$$= \frac{6.7 \times 10^{-11} \times 50 \times 50}{0.75^2}$$

$$= 3.0 \times 10^{-7} \text{N}$$

Exercise 7.3

An object of mass 1kg is 1000km above the surface of the Earth. Find the force of attraction between it and the Earth if the mass of the Earth is 6.0×10^{24}kg and the radius of the Earth is 6.4×10^6m. (Take G as $6.67 \times 10^{-11} \text{m}^3 \text{ kg}^{-1} \text{s}^{-2}$.) This answer has another name, what is it?

Forces which maintain equilibrium

Questions which equate the force needed to keep a body moving in a circle and either gravitational forces or electric forces are fairly common in examinations. If we consider a satellite moving round the Earth in a circular path, it needs a centripetal force to keep it in its circular path. This centripetal force is provided by the gravitational attraction between the satellite and the Earth. This provides us with the relationship:

$$\frac{mv^2}{r} = \frac{GMm}{r^2}$$

where M is the mass of the Earth, m is the mass of the satellite, v is the velocity of the satellite, G is the universal gravitational constant and r is the distance between their centres. Alternatively the centripetal force could be given as $F = mr\omega^2$.

Example

A satellite of mass m is placed in an orbit at a constant distance of 7.5×10^6m from the centre of a planet of mass 6.0×10^{24}kg. If G is 6.7×10^{-11}Nm^{2}kg^{-2}, find the speed at which the satellite must be travelling.

$$\frac{mv^2}{r} = \frac{GMm}{r^2}$$

We can see that $\frac{m}{r}$ can be cancelled from each side of the equation.

$$v^2 = \frac{GM}{r}$$

$$v = \sqrt{\frac{GM}{r}}$$

$$= \sqrt{\frac{6.7 \times 10^{-11} \times 6.0 \times 10^{24}}{7.5 \times 10^6}}$$

$$7320 \text{ms}^{-2}$$

Exercise 7.4

Satellites which transmit television signals are in geostationary or synchronous orbit which means that they always stay in the same position above the surface of the Earth. Find the height of the satellite if the mass of the Earth is 6.0×10^{24}kg, the radius is 6400km and G is 6.7×10^{-11} Nm^{2}kg^{-2}.

Magnetic forces

The third in the set of these rather strange forces is **magnetic force**. The phenomenon is similar to the other two in that bodies have an effect on each other from a distance without any visible connection but there the similarity ends. The formulae to be understood are completely different and the field characteristics are not the same. In its simplest form we can feel the repulsion between two like magnetic poles and the attraction between two unlike poles. Indeed the Earth itself behaves as if it has a large magnet in its core with the South pole of the magnet near magnetic North and the North Pole of the magnet near magnetic South. This is why the North-seeking pole of a magnet or compass needle is attracted towards the magnetic North pole of the Earth, rather than being repelled from it.

Electromagnetic forces on conductors

A conductor carrying a current at right angles to a magnetic field experiences a force. If the conductor is not perpendicular to the magnetic field, the force is smaller and if the angle between the lines of force in the field and the conductor is zero, there is no force. The direction of the force can be predicted using Fleming's left hand rule. This states that:

If the first finger, second finger and thumb of the left hand are held at right angles to each other and the First finger is pointing in the direction of the magnetic Field, the seCond finger in the direction of the Current then the thuMb points in the direction of the Movement.

The size of this force on a conductor is given by:

$$F = BIL\sin\theta$$

Where B is the magnetic field strength which is measured in teslas (T), I is the electric current in the conductor in amperes, L is the length of the conductor in the magnetic field in metres and θ is the angle between the conductor and the lines of force in the magnetic field.

If θ is 90°, sinθ is one so the force is maximum. If the conductor is parallel to the lines of force, sinθ is 0 and the force is zero.

Examples

A straight wire is carrying a current of 4.0 A. For 23cm of its length it is in a field of 5.0×10^{-3} T. Find the force on the wire if it is (a) perpendicular to the lines of force in the magnetic field and (b) if it is at an angle of 50° to the field lines.

a)
$$F = BIL \sin\theta$$
$$= 5.0 \times 10^{-3} \times 4.0 \times 0.23 \times \sin 90$$
$$= 4.6 \times 10^{-3} N.$$

b)
$$F = BIL \sin\theta$$
$$= 5.0 \times 10^{-3} \times 4.0 \times 0.23 \times \sin 50$$
$$= 3.5 \times 10^{-3} N.$$

Electromagnetic forces on particles

The section above dealt with charge moving through a magnetic field using a conductor as the medium through which it moves. It is possible for charged particles to move without a medium so long as they are moving through a vacuum and will therefore not be absorbed by the surrounding air. Again, for maximum force, the particles must be travelling at right angles to the field and if they are travelling parallel to the field the force is zero. The direction of the force is again given by Fleming's left hand rule but you must remember that the direction of the current is the direction of movement of positive charge. If it is a stream of electrons (negatively charged) the seCond finger is pointed in the direction from which the charge is coming. The formula for the magnitude of the force is:

$$F = Bqv\sin\theta$$

Where B is the magnetic field strength in teslas, q is the charge on each particle in coulombs, v is the velocity of the particles in ms^{-1} and θ is the angle between the direction of movement of the particles and the direction of the lines of force of the magnetic field. The path of the particles will be circular since the force is at right angles to the direction of travel.

Example
A stream of electrons pass across a magnetic field of strength 2.5×10^{-3}T at right angles to the lines of force and at a speed of 2.1×10^7ms^{-1}. Find the force which acts on each electron at right angles to its direction of travel. ($e = 1.6 \times 10^{-19}$C).

$$F = Bqv\sin\theta$$
$$= 2.5 \times 10^{-3} \times 1.6 \times 10^{-19} \times 2.1 \times 10^7 \sin 90$$
$$= 8.4 \times 10^{-15}\text{N}$$

Force on a coil in a magnetic field
An electric motor or a moving coil meter depends on the force exerted on a coil carrying a current in a magnetic field. Since there is an equal force on each side causing the coil to rotate in the same direction, we can say that a couple causes the rotation. The moment or torque of this couple is given by $BANI$ and the unit will be Nm. You may like to see if you can derive this formula. Remember it is only the sides of the coil which are parallel to the axis of rotation which will experience a force. This pair of forces will produce a couple for which you can calculate the torque. At some stage you will end up with a width × breadth of the coil which can be replaced by A, its area.

Exercise 7.5
A rectangular coil of 150 turns and with dimensions 8.0cm by 3.0cm is suspended vertically in a magnetic field of strength 8.0×10^{-3}T. If a current of 0.20A is passing through it, calculate the torque on the coil.

What we should have learnt from this chapter
▶ An electric field is a region where a body experiences a force due to the electric charge that it carries.
▶ The direction of the field is the direction in which a positively charged particle would move.
▶ Electric field strength is measured in newtons per coulomb.
▶ Electric field strength can also be given in volts per metre.

▶ Where two charged bodies exist, there will be a force between them.

▶ If the charges are the same the force will be repulsive, if different the force will be attractive.

▶ Coulombs law gives the force between two bodies as

$$F = \frac{1}{4\pi\varepsilon_0} \frac{Q_1 Q_2}{r^2}$$

▶ ε_0 is a universal constant known as the permittivity of free space.

▶ The electric field strength at a distance r from the centre of a charged body in a radial electric field is given by

$$E = \frac{1}{4\pi\varepsilon_0} \frac{Q}{r^2}$$

▶ In a gravitational field it is the mass of the body, rather than its charge which controls the size of the force.

▶ Newton's law of gravitation gives the force between two bodies as $F = G \dfrac{m_1 m_2}{r^2}$

▶ G is the universal gravitational constant.

▶ Gravitational field strength is the force per unit mass in the field and can be calculated for a radial field using

$$g = G \frac{M}{r^2}$$ where M is the mass of the body causing the field.

▶ Only forces of attraction are found in gravitational fields, no repulsion.

▶ A satellite stays in orbit because the gravitational force between the satellite and the planet provides the centripetal force necessary to keep it moving in a circle.

▶ A conductor carrying an electric current at right angles to a magnet field experiences a force.

▶ The direction of this force can be determined using Fleming's left hand rule.

▶ The size of the force can be found using $F = BIL\sin\theta$.

▶ Tesla (T) is the unit used to measure magnetic field strength.

▶ A charged particle or stream of charged particles travelling at right angles across a magnetic field will experience a force.

▶ The size of the force is given by $F = Bqv\sin\theta$.

▶ The direction of the force is given by Fleming's left hand rule. (Remember that the current direction is the direction of *positively* charged particles.)

▶ The path will be circular because the force is at right angles to the direction of travel.

▶ A coil in a magnetic field will experience a force given by $F = BANI$ where A is the area of the coil, N is the number of turns and I is the current in the coil.

Tutorial

Seminar discussion
Explain Fleming's left hand rule.

Study tips
You learn by interacting with the learning material and so you should try wherever possible to draw diagrams and charts that summarise information. You also have to work through the calculations yourself. It is not enough to simply read them.

8

Pressure As Force Per Unit Area

One-minute summary – In the study of pressure we see how larger forces can be supported or endured by spreading the force over a larger area so that each part of the body experiencing the force receives a smaller share. In this chapter we will:

▶ learn how we define pressure
▶ find out how to calculate the pressure exerted on a surface by an object
▶ understand the cause and effect of atmospheric pressure
▶ learn how to calculate the pressure beneath a column of water
▶ find out about the work which can be done by expanding gases and how this can be utilised.

Spreading the load

Spreading the load is a fairly well known concept. Skis spread the weight of the skier over the surface of the snow so that the skier does not sink in. Snowshoes do a similar job. Ice skates are the opposite, however, they deliberately concentrate the weight of the skier on a smaller area since the high pressure beneath the skate blade will cause the ice to melt briefly and provide a lubricant between skate and ice before refreezing again.

Pressure is force per unit area and is measured in **pascals** (Pa). One pascal is the same as one newton per square metre or one Nm^{-2}.

$$\text{pressure} = \frac{force}{area}$$

By keeping the blade of a knife sharp, all the force exerted on it is concentrated on a small area causing a high pressure. A similar principle applies with a drawing pin. The same force is exerted on your thumb when you push (the force is spread over a large area) but at the point of the drawing pin it is concentrated on a small area.

Pressure beneath the air

We are all familiar with the idea of atmospheric pressure. Our bodies have evolved to exist at standard atmospheric pressure. We know that atmospheric pressure varies from day to day and we can find this out from the weather charts on television. The pressure is caused by the column of air which stretches upwards from the surface of the Earth. If we know atmospheric pressure in pascals we know the weight of air acting on each square metre of the surface of the Earth.

Pressure beneath the water

A fish or a diver is under extra pressure due to the weight of the water. They are not crushed because the pressure inside their bodies also increases. The extra pressure caused by the water can be calculated using the formula for pressure due to a fluid:

pressure = 9.81 × depth of the liquid × density of the liquid

The pressure will be in pascals if the depth is in metres and the density is in kgm^{-3}.

Example

A giant ray is lying at the bottom of the ocean where it is 31.0m deep. If the area of the top surface of the ray is $0.210m^2$ and the sea water has a density of $1050kgm^{-3}$, find the force exerted on the upper surface by the water alone.

$$\begin{aligned} \text{pressure} &= 9.81 \times \text{depth} \times \text{density} \\ &= 9.81 \times 1.0 \times 1050 \\ &= 319315.5Pa \end{aligned}$$

$$\begin{aligned}
\text{force} \ &= \ \text{pressure} \times \text{area} \\
&= \ 319315.5 \times 0.210 \\
&= \ 67056.255 \\
&= \ 67000\text{Pa} \\
&= \ 67\text{kPa}
\end{aligned}$$

Work done by expanding gases

The energy required to drive a car along is acquired from an internal combustion engine. This energy results from the petrol and air mixture burning and expanding.

work done = pressure × change in volume

If a large change of volume is involved the pressure will get less as the volume change occurs. This pressure exerts a force on the piston in the engine causing the drive shaft to rotate.

Hydraulics

Hydraulic systems are those where a force is transmitted from one place to another by a liquid. The application of a force in one place increases the pressure in the liquid and since a liquid cannot be compressed, this pressure is transmitted in all directions through the liquid. This can act as a force multiplier since a small force on a small piston will produce a pressure in the liquid but if this pressure is applied to a larger piston, the force will be larger. Although we are increasing the force, the distance moved by the piston is reduced by the same proportion so that energy is conserved. This is another example of a force/distance trade off.

Example

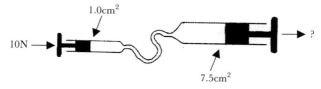

Fig. 19

The smaller syringe in Figure 19 is pushed in using a force of 10N. If we assume that the area of the piston is 1.0cm^2, the pressure is given by:

$$\text{pressure} = \frac{force}{area}$$

$$= \frac{10}{1.0 \times 10^{-4}}$$

$$= 1.0 \times 10^5 \text{Nm}^{-2}$$

At the larger piston, the area is 7.5cm^2.

$$\text{pressure} = \frac{force}{area}$$

$$\text{force} = \text{pressure} \times \text{area}$$

$$= 1.0 \times 10^5 \times 7.5 \times 10^{-4}$$

$$= 75\text{N}$$

What we should have learnt from this chapter

▶ Pressure is force per unit area.
▶ Pressure is measured in pascals which is the same as newtons per square metre.
▶ Pressure $= \dfrac{force}{area}$
▶ If the force is spread over a greater area the pressure is reduced.
▶ Atmospheric pressure is caused by the force of the air above

pushing down on the surface of the Earth.

▶ Pressure beneath a fluid is equal to 9.81 × depth of the fluid in metres × the density of the fluid.

▶ When a gas expands the work done is equal to pressure × change in volume.

▶ When a substance is a liquid, the particles are as close together as they can get unless very large forces are applied.

▶ This makes liquids incompressible, unlike gases which can have their particles pushed closer together. Liquids can therefore be used to transmit pressure and this is the basis of hydraulics.

Tutorial

Seminar discussion
Explain the difference between force and pressure.

Study tips
Be clear in your own mind about the relationship between pascals and newtons per square metre.

9

Forces Which Hold Particles Together

One-minute summary – It is possible to have fairly rigid bonds holding particles together or to have the particles able to move much more freely. In gases the particles are much further apart and it is these characteristics which provide the difference in the nature of the three states of matter. In this chapter we will:

▶ learn the difference between solids, liquids and gasses
▶ find out how a floor can support our weight
▶ understand why we feel lighter in water.

Solids, liquids and gases

In a solid substance the particles are as close together as they can get without other forces of repulsion being caused. A lattice of bonds joins the particles. If the bonds are very rigid the **solid** is hard to deform, but for some substances it is possible to stretch the bonds so that the substance stretches. When the stretching force is removed the substance returns to normal because the bonds return to their normal length. This type of solid is said to be perfectly **elastic**. Some substances can be made to stretch but they do not return to their original shape. These are **plastic** or inelastic and their bonds have changed during the stretching process because the particles have slid over one another. Many substances are elastic when small forces are applied but reach an elastic limit after which they no longer return to their original shape.

When a solid melts, energy is used to break the rigid bonds which hold the particles in fixed positions. This energy, which is called **latent heat**, allows the particles to move freely within the body of the **liquid** but they do not move further apart generally. This gives them a fixed volume but no fixed shape because they change their shape to fit the base of their container.

When a liquid boils or evaporates the particles acquire sufficient kinetic energy to be able to move freely without being held close together. This means that a **gas** has no fixed shape or volume.

Why don't we fall through the floor?

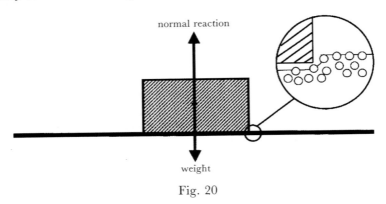

Fig. 20

Figure 20 shows a block standing on a surface. In order to support the block the surface must be able to exert an equal and opposite force to the weight of the block. Whether or not it can do this depends on how strongly the particles in the surface are held together.

When the block is placed on the surface it will deform the surface slightly and the amount of deformation will depend on the weight of the block and the size of the force needed to displace the particles in the surface. If the block is relatively heavy and the bonds between the particles in the surface are relatively weak, the bonds will be broken and the block will fall through the surface.

If the floor is to support us, our weight will displace the particles in the floor from their normal positions. The bonds between these particles will be strong enough to enable them to exert a force to prevent the particles being pulled apart completely. It is the combined effect of these forces trying to return the surface to its original state which causes the equal and opposite force to the weight as mentioned in Newton's third law.

What is Young's modulus?

Young's modulus concerns the stretching of thin samples of material such as wires. It is the ratio of stress to strain. The stress is the force per unit area of cross-section of the wire and the strain is the increase in length per unit length. Therefore we have:

$$\text{stress} = \frac{\text{force}}{\text{area}}$$

$$\text{strain} = \frac{\text{increase in length}}{\text{original length}}$$

$$\text{Young's modulus} = \frac{\text{stress}}{\text{strain}}$$

The units for Young's modulus are Nm^{-2}. The example below will demonstrate its use.

Example

A steel wire with a length of 2.5m and a diameter of 0.35mm is hung from a support. When a weight of 60N was hung on the wire, its length increased by 5.8mm. Find Young's modulus for the material from which the wire is made.

$$\text{area} = \pi r^2$$

$$= \pi \times (0.175 \times 10^{-3})^2$$

$$= 9.62112 \times 10^{-8} m^2$$

$$\text{stress} = \frac{\text{force}}{\text{area}}$$

$$= \frac{6}{9.62112 \times 10^{-8}}$$

$$= 6.2362 \times 10^8 Nm^{-2}$$

$$\text{strain} = \frac{\text{increase in length}}{\text{original length}}$$

$$= \frac{5.8 \times 10^{-3}}{2.5}$$

$$= 2.32 \times 10^{-3}$$

$$\text{Young's modulus} = \frac{\text{stress}}{\text{strain}}$$

$$= \frac{6.2362 \times 10^8}{2.32 \times 10^{-3}}$$

$$= 2.7 \times 10^{11} \text{Nm}^{-2}$$

Exercise 10.1

A wire of diameter 0.150 cm and 4.00m long is hung vertically with a load of 10.0kg on the end. Find the extension if Young's modulus for the wire is $2.00 \times 10^{11} \text{Nm}^{-2}$.

Getting lighter in water

We have all experienced the way in which a body feels lighter in water than it does when suspended in air. This is made use of in physiotherapy where patients recovering from problems with their limbs can exercise them with the water supporting part of their weight. Even expensive racehorses can be exercised in special baths which allow them to use their legs without them needing to carry their full weight. It was Archimedes (287–212BC), the Greek mathematician, who first investigated this phenomenon. After a well-known experience involving an overflowing bath he suggested that the apparent loss in weight was equal to the weight of water displaced by the object. We have therefore been left with Archimedes' principle which states that:

When a body is partially or totally immersed in a fluid it experiences an upthrust equal to the weight of fluid displaced.

The word **fluid** is used because the effect occurs in gases as well as liquids and can be used to solve problems dealing with balloons filled with helium or hot air.

Why do some objects float?

If the object which is immersed in the water is less dense than the water itself, the water exerts an upthrust equal to the weight of the object, even though it is not submerged. This means that the object will just sink until it experiences an upthrust equal to its own weight.

Ships are able to **float** although the density of the metal is greater than that of water. This is possible because there is a lot of air filled space in the ship although the bottom of the ship is solid. This means that it appears to have a large volume because it displaces a lot of water but has only a relatively small mass of metal. Alternatively you could argue that the average density of the ship is less than that of water.

Example

A rectangular picnic basket falls from a rowing boat while a family is on its way to an island in the middle of a lake. The dimensions of the top are 35cm × 45cm and the basket is 20cm deep. If the basket weighs 9.5kg and floats the correct way up, what depth is under the water? (Density of water in the lake is 1000kgm^{-1}.)

The weight of the basket is $mg = 9.5 \times 10 = 95\text{N}$. If the basket is to float the weight of water displaced must also be 95N.

$$\text{mass of water displaced} = 9.5\text{kg}$$

$$\text{density} = \frac{mass}{volume}$$

$$\text{volume of water displaced} = \frac{mass}{density}$$

$$= \frac{9.5}{1000}$$

$$= 9.5 \times 10^{-3} \text{ m}^3$$

Let us call the depth of the basket under water d.

$$\text{volume under water} = 0.35 \times 0.45 \times d$$
$$= 0.1575d \text{ m}^3$$

$$d = \frac{9.5 \times 10^{-3}}{0.1575}$$

$$= 0.0603\text{m}$$
$$= 6.0\text{cm.}$$

Example

A balloon of volume 180m^3 is filled with helium of density 0.18kgm^{-3}. The mass of the pilot and basket is 81kg and the mass of the fabric is 76.3kg. If the surrounding air has a density of 1.19kgm^{-3}, what is the maximum additional load which the balloon can lift? (Assume that volume of the basket and contents is negligible compared with that of the balloon.)

$$\text{volume of air displaced by the balloon} = 180\text{m}^3$$

$$\text{mass of air displaced by the balloon} = \text{volume} \times \text{density}$$

$$= 180 \times 1.19$$

$$= 214.2\text{kg}$$

If the balloon is to float, the downward force must equal the upward force.

$$\text{total mass of the balloon cannot exceed } 214.2\text{kg}$$

$$\text{additional load} = 214.2 - 81 - 76.3 - (180 \times 0.18)$$

$$= 24.5\text{kg}$$

Exercises

10.2 A watertight container has dimensions 5m × 3m × 4m and is being unloaded from a ship by a crane. The tension in the cable of the crane is $1.5 \times 10^6\text{N}$. As a result of a fault in the lifting mechanism, the container slowly falls towards the water, while still attached to the cable. What is the tension in the cable when the container is
(a) half submerged
(b) completely submerged
(c) resting on the bottom?

10.3 A large metal key is attached to the end of a spiral spring which obeys Hooke's law. The length of the spring is 25.0cm before the metal key of mass 750g is hung on it and 31.2cm while it is hanging. When the key is totally immersed in water of density 1000kgm^{-3} the length of the spring is 27.5cm. Find the volume of the metal key.

What we should have learnt from this chapter

▶ In solid substances the particles are held close together by fairly rigid bonds.

▶ The nature of these bonds decides whether the solid is elastic or inelastic.

▶ A solid has a fixed shape and a fixed volume.

▶ In liquids the particles are still held closely together but they are free to move within the body of the liquid.

▶ A liquid has fixed volume but no fixed shape.

▶ The particles of a gas are not held together in any way.

▶ A gas has no fixed shape or volume.

▶ A solid surface is able to support a weight if the bonds exert a force when the particles are displaced and combined effect of the forces from the displaced particles is sufficient to provide a normal reaction equal and opposite to the weight of the object.

▶ Young's modulus is concerned with the stretching of thin wires.

▶ $\text{Stress} = \dfrac{\text{force}}{\text{area}}$

▶ $\text{Strain} = \dfrac{\text{increase in length}}{\text{orginal length}}$

▶ $\text{Young's modulus} = \dfrac{\text{stress}}{\text{strain}}$

▶ Objects experience an upthrust when partly or totally immersed in a fluid.

▶ According to Archimedes this upthrust is equal to the weight of fluid displaced.

▶ Objects float when they can displace sufficient weight of fluid to equal their own weight.

Tutorial

Seminar discussion

Explain why ships float.

Study tips

Examine the learning outcomes from this chapter and consider putting them on flash cards. These are cards which simply have one statement on them that you can use as a reminder. Make sure you use colours and images. The point you need to understand is that it is the act of creating the cards that is the learning experience. This is because it is creative and creativity is a whole brain function.

10

Forces Which Cause Oscillations

One-minute summary – Objects such as a pendulum or a mass on a spring exhibit simple harmonic motion and this types of motion has a well defined pattern. In this chapter we will:

▶ understand the characteristics of simple harmonic motion
▶ discover other examples
▶ learn how to assess other possible examples.

Restoring forces

Fig. 21

Figure 21 shows a mass hanging on a spring. When at rest it is in **equilibrium** because the weight of the mass acts downwards, but there is an **upward force equal** and **opposite** to this which results from the stretching of the spring. As mentioned in Chapter 1 the stretched spring tries to return to its original state hence there is a restoring force which opposes the weight of the mass.

If the spring is pulled down further and then released it will **oscillate** up and down. This is because there is now a **restoring force** greater than the weight of the mass so it pulls the mass upwards. It will pass through the equilibrium position moving at speed and will then be decelerated because the weight is now greater than the upward force. The restoring force will always be proportional to the displacement and this is the requirement for simple harmonic motion. The features of **simple harmonic motion** are:

1. The restoring force and therefore the acceleration are maximum when the displacement is maximum.

2. The restoring force and therefore the acceleration are zero when it passes through the equilibrium position i.e. zero displacement.

3. The velocity is maximum when it passes through zero displacement.

4. The velocity is zero at maximum displacement because it is changing direction.

Further examples of simple harmonic motion

Any oscillating system where the restoring force is proportional to the displacement will display simple harmonic motion. These include:

▶ A simple pendulum since as it swings back and forth its restoring force will be proportional to the height above the rest position. It accelerates as it falls down one side and then decelerates as it passes up the other side until it comes to rest and begins to accelerate downwards again.

▶ A ball bearing rolling on a concave surface is just like the pendulum bob without the string.

▶ Other examples include the canvas on a trampoline, the plucked string on a guitar and the skin on a drum.

What we should have learnt from this chapter

▶ Many oscillating objects demonstrate simple harmonic motion.

▶ For an object exhibiting simple harmonic motion, the restoring force is proportional to the displacement.

Tutorial

Seminar discussion
Using PowerPoint, create a multimedia slide show to explain simple harmonic motion.

Study tips
Use the Internet to research Galileo's work with pendulums.

Forces Which Allow Us To Fly

One-minute summary – The **Bernoulli effect** is the result of a fluid such as air speeding up as it passes through a narrow space in order that the overall flow is maintained. This causes a reduction in pressure in the narrow opening. An adaptation to this causes flight. We will:

► learn more about the Bernoulli effect
► see how this effect provides lift on an aeroplane wing
► briefly study the forces on a fixed wing aircraft in flight
► think about how this affects a helicopter.

The Bernoulli effect

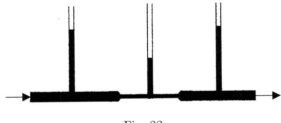

Fig. 22

Figure 22 shows a liquid flowing through a tube which has a narrow section or venturi as it is sometimes called. The level of the fluid in the vertical tubes shows that the pressure in the narrow section is lower than that in the wider section. The liquid flowing more **quickly** through that section causes this **reduction** in pressure. This is similar to a rocky ledge across the bed of a river causing rapids because the water flows more quickly in order to get through the narrower space above the rock. A similar effect is found when a gas is moving along a tube so the phenomenon applies to fluids in general. The effect is seen in a carburettor where air passing through a narrow venturi causes a reduction in pressure allowing petrol to be force into the air stream. Paint sprayers work on a similar principle.

The flow of air over an aerofoil

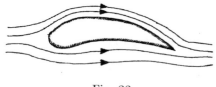

Fig. 23

Figure 23 shows the flow of air over an aircraft wing. The wing is curved at the front (the leading edge) and more pointed at the back (the trailing edge). This means that the distance the air has to travel over the wing is greater than the distance which it travels beneath the wing. This means that the air **flows faster over** the wing than under the wing. The air pressure above the wing is therefore less than atmospheric pressure whereas beneath the wing it is equal to atmospheric pressure. It is this pressure difference which **causes lift** and therefore allows an aeroplane to fly.

Lift, weight, thrust and drag

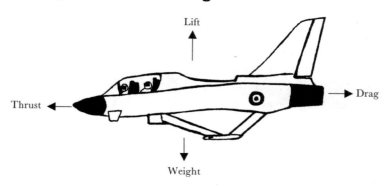

Fig. 24

Figure 24 shows the forces which act on an aircraft in flight. The vertical forces are the weight of the aircraft and the lift provided by the flow of air over the wings. The horizontal forces are the **thrust** provided by the engines and the **drag** caused by the friction between the body and the air. These two sets of forces are not independent, however because the **lift** is dependent on the speed of the airflow over the wings which is dependent on the speed of the aircraft and hence on the thrust provided by the engines. The control of the aircraft is gained by the use of the ailerons on the trailing edges of the wings and the elevators and rudder on the tail plane. These allow the pilot to change the direction of the lift, thrust and drag, but not, of course, the weight which always acts downwards. Note that lift and **weight** do not pass through the same point. The result of this is a turning moment which is balanced by the pilot using the controls.

And what about helicopters?

The blades on the main rotor are just like thin aeroplane wings which provide lift as they rotate. The shape is very similar to that of an aeroplane wing. The reason for the additional vertical rotor on the tail is to prevent rotation of the helicopter. If the blades of the main rotor rotate one way, there should be an equal and opposite force causing the body to rotate the other way. The small vertical rotor counteracts this and holds the body stationary.

What we should have learnt from this chapter
▶ When a fluid flows through a narrow section in a pipe a reduction in pressure results from the fact that the fluid is travelling more quickly.
▶ The lift on an aircraft wing results from the pressure reduction caused by air flowing faster over the top of the wing.
▶ Aircraft stability is dependent on equilibrium between lift, weight, thrust and drag.

Tutorial

Seminar discussion
Explain the Bernoulli effect with diagrams.

Study tips
Draw the rotor blades on a helicopter and show the forces that keep the aircraft stable.

12

Useful Web Sites

Introduction

In recent years the Internet has become a very useful source of free information. It is important to remember that there is no control which can be placed on the sites so incorrect information will not be identified. It is therefore very important that you should be aware of the source of the information. A site set up by a university or professional organisation is more likely to contain correct information than a private one. Many of these sites listed are private ones. They all seem to be providing good information but they are not static and the information could be changed before this book is published, or the site could be removed completely.

The university sites are often more academic and supply information above the level required for our purpose.

In short, you must be wary of a source and look for other supporting evidence. It is little different from the problems associated with secondary historical sources which could either be biased or just incorrect.

Some of the sites provide a complete physics course and you could use them as a valuable additional source since a different person's viewpoint can often clarify a concept. Other sites are just a glossary of terms and some just deal with applications of individual principles. The remaining sites are just for entertainment (which may or may not be to your taste).

Learn Physics Today
http://library.advanced.org/10796/

This is an online physics tutorial which provides a mini physics course. It does not just deal with forces but includes many basic areas of physics. It is useful to see the laws stated in different forms. It can help to emphasise that the exact wording of a law is not critical since they are seldom the exact words used by the original scientist. In fact many of them have been translated from a different language. The important thing is to get the meaning exactly correct and to include any constraints which may be needed to fix the experimental conditions such as 'so long as the elastic limit has not been exceeded' or 'as long as the temperature remains constant'.

Chapter 4, 'Forces and Newton's Laws' covers Newton's laws of motion as its name suggests and Chapter 5, 'Motion in Two Dimensions', includes vector addition and resolution of forces. Circular motion is in Chapter 6 (Projectile and Periodic Motion) and Coulomb's law, electric fields and gravitational fields are in Chapters 11 and 12.

S-cool

http://www.s-cool.co.uk
This is a British site which covers a number of subjects at A level and each subject has a useful revision site.

Funderstanding

http://www.funderstanding.com/k12/coaster/
This is one of several sites connected with a roller coaster. Although not specifically connected to forces it provides a lot of fun by allowing you to design your own roller coaster. If you get the first hill too high compared with the second hill your car flies off the track. If you get it too low the car will not reach the top of the second hill and it then provides us with an example of simple harmonic motion. The final hurdle is to get the car to go round a loop.

The Physics of Amusement Parks

http://library.thinkquest.org/2745/data/openpark.htm
This is another amusement park site. It allows you to choose an application of a principle of physics and it also provides an explanation of the principle as well as the application.

The Physics Classroom

http://glenbrook/k12.il.us/gbssci/phys/phys.html
This is a large site which provides a complete physics course.

How to Study Physics

http://wwwrel.ph.utexas.edu/~larry/how/how.html

As its name suggests, this site is more concerned with the techniques of learning physics rather than actual physics content.

Physics for Beginners

http://physics.webplasma.com/physicstoc.html

This starts at a basic level but students should not be put off by its name. It does cover a great deal of the groundwork and it is ideal for any student who needs to start studying a topic from scratch.

Online Mechanics Course

http://www.mcasco.com/p1intro.html

This leads to an online mechanics course by an American named J D Jones. It gives links to an alphabetical list of word definition, rather like Chapter 13 of this book. It covers a range of topics from Newton's laws to circular motion. The approach is fairly descriptive at the start of each topic but involves quite a lot of mathematics after this. Pure mathematicians would probably find this site quite interesting.

How Stuff Works

http://www.howstuffworks.com/index.htm

This is the homepage for a fairly extensive web of scientific explanations for everyday devices or events. There is even a site which explains the workings of the web site. The most relevant places to visit in connection with forces are:

▶ How a helium balloon works at *http://www.howstuffworks.com/helium.htm*

▶ How airplanes work at *http://www.howstuffworks.com/airplane1.htm*

▶ How helicopters work at *http://www.howstuffworks.com/helicopter.htm*

▶ How hydraulic machines work at *http://www.howstuffworks.com/hydraulic.htm*

▶ How a block and tackle and other force/distance tradeoffs work at *http://www.howstuffworks.com/pulley.htm*

The Physics Factory

http://fp.physics.f9.co.uk

This site is produced by Richard Smith who is IT co-ordinator and formerly head of physics at Fullbrook School in Surrey. It includes a very useful web site search which will take you to any topic which you choose from the extensive range which is available. There is also a useful glossary of terms.

About.com

http://physics.about.com/science/physics

This site provides a gateway to many other sites featuring physics topics.

Eric's Treasure Trove of Physics

http://www.astro.virginia.edu/~eww6n/physics

This is yet another site which offers a glossary of physics terms. It would be a useful exercise to compare the definitions which are given on the different sites (and the ones given in Chapter 13 of this book).

Physics Humor Links

http://www.escape.ca/~dcc/phys/humor.html

From the spelling you can see that this is obviously American humour (or humor).

The Physics Humor Page

http://www.servtech.com/public/wkimler/humor/humor.html

American humor again. My computer doesn't like it because it keeps putting a red squiggly under the word humor every time I type it. Are you amused by it?

Scientific American

http://www.sciam.com/askexpert/

This site allows you to put a question on a scientific topic to experts. The most useful section is probably the archive section which allows you to view answers to questions which have been asked previously.

Physics Lecture Demonstrations

http://www.mip.berkeley.edu/physics/physics.html

This site gives instructions for all the physics demonstrations which can be given during lectures at the University of California at Berkeley.

13

Definitions and Laws

Definitions

Acceleration – Acceleration is the rate of change of velocity.

Aerofoil – The shape of the cross-section of an aeroplane wing – curved at the leading edge and pointed at the back.

Banking – This is the process of raising a road or track to produce a force towards the centre of the curve when a vehicle is turning a corner.

Bernoulli effect – This is the result of a fluid speeding up as it passes through a narrow space in order that the overall flow is maintained. It causes a reduction of pressure in the narrow opening.

Centre of gravity – The centre of gravity of a body is the point about which the whole weight of the body may be considered to act.

Centripetal force – The centripetal force is the force towards the centre which keeps a body travelling in a circle.

Couple – A couple is a pair of equal turning forces acting on opposite sides of an object and supporting rotation in the same direction.

Displacement – Displacement is distance moved in a given direction. It is the vector equivalent of distance.

Drag – The frictional force on a body as it moves through a fluid.

Efficiency – This compares the work done with the energy supplied. Efficiency = useful work done by the system/total energy supplied to the system.

Elastic – A process or collision is said to be perfectly elastic if the energy distribution after the process or collision is exactly the same as before it.

Elastic limit – The elastic limit of a spring is the point after which a spiral spring no longer returns to its original length after the force is removed.

Electric field – An electric field is a region where a charged body experiences a force due the charge which it is carrying.

Electric field strength – The electric field strength at a point is the force exerted by the field on a body carrying unit positive charge.

Energy – Energy is the capacity to do work.

Equilibrium – When a body is in equilibrium there must be no resultant linear force in any direction or any resulting turning force.

Fluid – A fluid is a liquid or a gas.

Force – A force is a push or a pull.

Force multiplier – A force multiplier is a device which allows us to use a small force in order to exert a larger force.

Friction – Friction is a force which opposes motion.

Gas – A gas has neither fixed volume nor fixed shape.

Gravitational field – This is a region where a body experiences a force due to its mass.

Gravitational force – This is a force exerted on a body due to its mass.

Hydraulics – A hydraulic system is one where a force is transmitted through the pressure in a liquid.

Impulse – Impulse is equal to the change in momentum. It is also equal to the product of accelerating force and accelerating time.

Inertia – Inertia is a measure of the resistance which an object has to a change in velocity.

Joule – The unit of work and energy. It is equal to the work done when a force of one newton moves its point of application by one metre.

Lift – The upward force on a wing due to the reduced pressure above it.

Liquid – A liquid has fixed volume but no fixed shape.

Mass – Mass is a measure of the amount of matter contained in a body. It is related to the number and type of atoms.

Mechanical advantage – Mechanical advantage $= \dfrac{\text{load}}{\text{effort}}$

Moment – The moment of a force is the product of the size of the force and the perpendicular distance from its point of application to the turning point.

Newton – One newton is the force which will cause a body of mass one kilogram to have an acceleration of one metre per second per second ($1\,\text{ms}^{-2}$).

Newtonmeter – A newtonmeter is used to measure forces. It can also be called a forcemeter.

Pascal – The pascal is the unit of pressure and is equal to one newton per square metre.

Plastic – A process or collision is plastic if the energy distribution is not the same after the process or collision as it was before.

Power – Power is the rate of doing work or power = work done/time taken

Pressure – Pressure is force per unit area.

Reaction – A normal reaction is the force exerted by a surface in response to an applied force. The reaction is always perpendicular to the surface and equal to the component of the applied force which acts in the opposite direction.

Restoring force – This is the force which tends to move a body back to its rest position after it has been displaced.

Simple harmonic motion – An oscillating body is undergoing simple harmonic motion if the restoring force is proportional to the displacement from the rest position.

Solid – A solid has fixed shape and fixed volume.

Speed – The rate of change of distance with time. It is not a vector.

Spring constant – The spring constant is the force needed to stretch the spring by unit length.

Strain – This is given by change in length divided by original length.

Stress – This is force divided by cross-sectional area.

Tension – The tension in a string or rod is the force which is acting along it.

Terminal velocity – A body is travelling at its terminal velocity when the resistive forces which oppose motion are equal to the accelerating forces which are causing motion. No resultant acceleration occurs.

Tesla – This is the unit of magnetic field strength. A magnetic field has a strength of one tesla if a conductor of length one metre carrying a current of one ampere at right angles to the magnetic field experiences a force of one newton.

Thrust – The forward force produced by an engine such as in an aircraft.

Torque – A torque is the moment of a couple. It is equal to one of the forces multiplied by the distance between them.

Upthrust – This is the upward force exerted on a body due to immersion in a fluid.

Vector – Vectors are quantities which have direction as well as size.

Viscosity – Viscosity is friction in fluids.

Velocity – Velocity is the rate of change of displacement with time (in a given direction).

Watt – This is the unit of power and is equal to one joule of energy being transferred per second.

Weight – Weight is the force of gravity which acts on a body.

Work done – Work done = force × distance moved in the direction of the force.

Young's modulus – Young's modulus = $\dfrac{\text{stress}}{\text{strain}}$

Laws

Archimedes' principle – When a body is partially or totally immersed in a fluid it experiences an upthrust equal to the weight of fluid displaced.

Coulomb's law – Coulomb's law states that the force between two charged bodies is proportional to the product of the charges and inversely proportional to the square of the distance between their centres.

Hooke's law – The extension of a spiral spring is proportional to the force applied.

Fleming's left hand rule – If the first finger, second finger and thumb of the left hand are held at right angles to each other and the First finger is pointing in the direction of the magnetic Field, the seCond finger in the direction of the Current then the thuMb points in the direction of the Motion.

Newton's first law – A body stays at rest or travelling with constant velocity unless acted upon by an external force.

Newton's second law – The acceleration of a body is proportional to the applied force and inversely proportional to its mass.

Newton's third law – Action and reaction are equal and opposite.

Newton's law of gravitation – Newton's law of gravitation states that the force between two bodies is proportional to the product of their mass and inversely proportional to the square of the distance between them.

Principle of moments – For a body in equilibrium, the sum of the anticlockwise moments is equal to the sum of the clockwise moments about the same point.

Answers to Exercises

Chapter 1
1.1. Force constant is $240 Nm^{-1}$ and the elastic limit is passed at a force of 7.0N.
1.2. 0.15m.

Chapter 2
2.1. 38N at an angle of 13° from the 25N force.
2.2. 72N at an angle of 34° from the 60N force.
2.3. 52N at an angle of 35° from the 60N force.
2.4. Horizontal force is 577N and the tension in the displaced chain is 1150N.

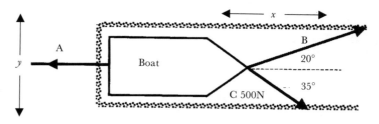

Fig. 25

2.5. Angle is 29° and the weight of the crate is 1900N.

Chapter 3
3.1. $2.6 \times 10^4 N$.
3.2. a) $3 \times 10^4 N$.
    b) $1 \times 10^4 N$.
    c) $1 \times 10^4 N$.
3.3. $2.5 \times 10^5 N$.
3.4. 530N.

Chapter 4
4.1. 1.7m from the bottle of drink.
4.2. 500N.
4.3. 2.6m from the tool bag.
4.4. Reaction at A is 52000N, Reaction at B is 43000N.
4.5. 55N.

Chapter 5
5.1. 240W.
5.2. 620N.
5.3. 0.29m.

Chapter 6
6.1. $13ms^{-1}$ Doubling the mass would have no effect.
6.2. 55N at the top and 66N at the bottom.
6.3. The minimum will be at the top of the loop since the pilot's weight will provide part of the centripetal force. Resultant force holding the pilot in their seat will be 750-707 = 43N. At the bottom of the loop it will be 750 + 707 = 1457N or 1460N.
6.4. 0.32 revolutions per second.
6.5. $26.5ms^{-1}$
6.6. $11°$.

Chapter 7
7.1. 1.0×10^{-11}J.
7.2. 9.2×10^{-8}N.
7.3. 7.3N It is the gravitational field strength at that point.
7.4. 36,000km (Use $mr\omega^2 = \dfrac{GMm}{r^2}$).

7.5. 5.8×10^{-4}Nm.

Chapter 10
10.1. 1.13×10^{-3}m.
10.2. (a) 1.2×10^{6}N.
      (b) 0.9×10^{6}N.
      (c) zero.
10.3. 4.5×10^{-4}m^{3}.

Index